Satellites of the Outer Planets

Satellites of the Outer Planets

Worlds in their own right

DAVID A. ROTHERY

Lecturer in Earth Sciences
The Open University

To Catriona

Wishing you good luck with S207

Dave Rothery

Lancaster 1995

CLARENDON PRESS · OXFORD

Oxford University Press, Walton Street, Oxford OX2 6DP
Oxford New York Toronto
Delhi Bombay Calcutta Madras Karachi
Kuala Lumpur Singapore Hong Kong Tokyo
Nairobi Dar es Salaam Cape Town
Melbourne Auckland Madrid
and associated companies in
Berlin Ibadan

Oxford is a trade mark of Oxford University Press

Published in the United States
by Oxford University Press Inc., New York

A catalogue record for this book is available from the British Library

Library of Congress Cataloging in Publication Data
Rothery, David A.
Satellites of the outer planets: worlds in their own right/
David A. Rothery.
Includes bibliographical references and index.
1. Satellites—Outer planets. I. Title.
QB401.R67   1992   523.4—dc20   91–27437 CIP
ISBN 0 19 854289 5 (Hbk)
ISBN 0 19 854290 9 (Pbk)

Printed in Hong Kong

To Edwin and Elanor

Preface

Time was when, very properly, geologists confined their attention to the Earth. After all, its profusion of rivers and mountains, beaches and cliffs, and above all the fact that we live here, are what enabled the science of geology to get going. Geology is based on observations made on the Earth. Concepts such as the principal of superposition (younger deposits tend to overlie older deposits) and uniformitarianism (the present is the key to the past) are all 'home-grown'.

During the latter part of this century, the earth sciences were revolutionized by the development of techniques and the discovery of processes that have given us insights into how the Earth works as a whole. Plate tectonics is the obvious example of this. At the same time, however, data were beginning to accumulate, at first very slowly but then increasingly rapidly, that mean we no longer have just one planet to explore and to seek to understand. Serious geological studies of the Moon began in the early 1960s, first by telescopic observations, then using high-resolution cameras in lunar orbit, and soon by manned landings and the return of samples. Space probes to other planets, notably Mars, began to reveal evidence of volcanism, faulting, erosion, and transport by wind and water that were at once familiar and yet subtly different from what we are used to at home.

The other terrestrial planets, Mercury, Venus, the Moon, and Mars, so-called because they resemble the Earth in having a rocky outer layer on which their geological histories are written, have received a fair amount of attention. There are now several good texts on comparative planetology that deal with these worlds in detail, although there has been a curious and unfortunate reluctance within much of the earth science community to read them.

Here I want to look at a different class of bodies, the moons of the outer planets: Jupiter, Saturn, Uranus, and Neptune. Most of these satellites lack a rocky crust, but are none the less surfaced by hard, rigid material—ice of various kinds—that behaves in almost all respects like rock. They constitute about a quarter of the mappable terrain in the solar system, only slightly less than the total surface area of the Earth (including the ocean floor). These worlds are important objects for study if geologists are to gain the most that comparative planetology can offer in the way of insights into how the Earth and other planet-sized bodies have evolved, and what processes may occur on and beneath their surfaces.

To the astronomer, the moons of the outer planets used to be convenient points of light: markers in the game of celestial mechanics that were useful for measuring the mass of the planet round which they orbit, or distant clocks for determining the speed of light. Now, thanks to the outstanding achievements of the two *Voyager* space probes, we have been privileged to see their surfaces from close by, and can recognize them for what they truly are: worlds in their own right on which the forces of geology have left their marks in a variety of related and yet intriguingly varied ways.

This book is not intended as an all-embracing treatise on the satellites of the outer planets. There are other works, noted in the Bibliography, which come much closer to fulfilling that role. Here the smaller satellites are discussed only in passing, and the larger ones are treated from a geological point of view. The matter of the origin of these bodies is covered only briefly, although I believe in sufficient depth for their significance in understanding the formation and evolution of the solar system to be appreciated.

As is standard practice these days, the term *billion* is used to mean a *thousand million* (10^9) and not a million million (10^{12}). I have attempted to avoid jargon as much as possible, to make this book accessible to a wide audience. I have tried to exlain terms that may be unfamiliar (or used in an unfamiliar sense) where they first appear, and have included a glossary of such terms at the end of the main text.

This book would not have been possible without the generous provision of illustrative material by several people, especially Alfred McEwen and Marion Rudnyk, or the efforts of my friends and colleagues who read parts of it or took the time to discuss various issues with me, especially Helen Rothery, and John Spencer, who read the whole thing in draft form. Nor could it have been written without the US taxpayer, whose government has funded the lion's share of mankind's exploration of the outer solar system. To all of these people, my thanks. The specific sources of the illustrative material used are listed in the figure and plate credits, on p. ix.

Pury End   D.A.R.
September 1991

As this book goes into its third impression, within two years of initial publication, I would like to take this opportunity to thank those friends, colleagues, and readers otherwise unknown to me who have taken the trouble to point out typographical and a few other errors. I am always open to suggestions as to how the book may be further improved.

Pury End   D.A.R.
December 1993

Acknowledgements

The spacecraft images used in this book all originated from NASA. Those that are not individually acknowledged below are either widely distributed standard product images or (in a few instances only) have been digitally processed by the author.

Figures

Fig. 1.8 National Space Science Data Center, World Data Center A for Rockets and Satellites (Smith, B. A.); **Fig. 2.3** modified after Consolmagno, G. J. and Lewis, J. S. (1978) *Icarus*, **34**, 284; **Fig. 2.5** modified from Schubert, G., Spohn, T. and Reynolds, R. T. (1986) in *Satellites* (ed. Burns, J. A. and Matthews, M. S), p. 234, University of Arizona Press; **Fig. 3.2** NASA; **Figs 3.8**, **5.7(c)**, **5.19**, **5.23**, **5.25**, **6.1**, **6.15**, **6.16(a)**, **6.20**, **6.28**, **6.29(a)**, **7.2**, **7.34(a)**, U.S. Geological Survey, Flagstaff; **Fig. 4.3** modified from McKinnon, W. B. and Parmentier, E. M. (1986) in *Satellites* (ed. Burns, J. A. and Matthews, M. S.), 723, University of Arizona Press; **Fig. 4.5** modified from Stevenson, D. J. (1982) *Nature*, **298**, 144; **Fig. 5.12** National Space Science Data Center, World Data Center A for Rockets and Satellites (Kosofsky, L. J.); **Figs 5.15**, **7.10** J. R. Spencer; **Fig. 5.17** Ellsworth, K. and Schubert, G. (1983) *Icarus*, **54**, 503; **Fig. 5.20** *ibid.*, p. 501; **Figs 6.3**, **6.5**, **6.12** University of London Observatory; **Fig. 6.8** modified from Shoemaker, E. H., Lucchitta, B. K., Wilhelms, D. E., Plescia, J. B. and Squyres, S. W. (1982) in *Satellites of Jupiter* (ed. Morrison, D.), p. 513, University of Arizona Press; **Fig. 6.14** modified from Moore, J. M. (1984) *Icarus*, **59**, 213; **Fig. 6.17** modified from Consolmagno, G. J. (1985) *Icarus*, **64**, 410; **Fig. 6.22** modified from Plescia, J. B. (1987) *Nature*, **327**, 202; **Fig. 6.31** modified from Smith, B. A. *et al.* (1986) *Science*, **233**, 62; **Fig. 6.35** modified from Janes, D. M. and Melosh, H. J. (1988) *Journal of Geophysical Research*, **93**, 3132; **Fig. 6.36** modified after Croft, S. K. (1988) *Lunar and Planetary Science Conference*, **19**, 226; **Fig. 7.14** C. M. M. Oppenheimer; **Fig. 7.21** modified from Schenk, P. M. and McKinnon, W. B. (1989) *Icarus*, **79**, 80; **Fig. 7.24** modified from Morrison, D., Owen, T. and Soderblom, L. A. (1986) in *Satellites* (ed. Burns, J. A. and Matthews, M. S.), p. 781, University of Arizona Press; **Fig. 7.26** S. A. Drury; **Fig. 7.28** J. Edmond (MIT and Woods Hole Oceanographic Institution); **Fig. 7.29** modified from Croft, S. K. (1990) *Lunar and Planetary Science Conference*, **21**, 249; **Fig. 8.2** Space Telescope Science Institute.

Plates

Plates **3**, **6**, **7** U.S. Geological Survey, Flagstaff.

Contents

5 Dead worlds

1

Introduction

So many worlds, so much to do,
So little done, such things to be
Tennyson, *In Memoriam*

Galileo made one of his more remarkable discoveries during the winter of 1610 when he turned one of the earliest telescopes on to the planet Jupiter and saw that four smaller bodies were moving round it. Many philosophers of the time refused to accept the truth of this observation, because it undermined the long-held belief that the Earth was at the centre of the universe, and that all the heavenly bodies (the Sun included) must go round it.

For a while, Jupiter's galilean satellites, as they became known, played a major role in the advance of science through the information they gave about the universe around them. As Galileo realized, the fact that they do indeed go round the planet Jupiter was a strong argument in favour of the Copernican view of the cosmos and against the then established view, which had the Earth at the centre of everything. Sixty years or so later, after the periodic motions of these satellites had become sufficiently documented, the Danish astronomer Ole Rømer measured slight differences between the predicted and observed times at which each of these moons disappeared into the shadow of Jupiter. He correctly realized that the orbits are as regular as clockwork, but, because the distance between the Earth and Jupiter varies, so does the time taken for the light to reach the Earth. Thus Rømer was able to show that the speed of light must be finite, and in fact made a surprisingly accurate determination of its value. Soon, with the benefits of Newton's gravitational theory and laws of motion, the size and period of the orbits of Jupiter's satellites enabled the enormous mass of the planet itself to be calculated.

Once a family of satellites had been documented around Jupiter, it was comparatively unremarkable to discover that the other outer planets in the solar system also have moons. By 1700, five satellites of Saturn were known. Within six years of his discovery of the planet Uranus in 1781, Sir William Herschel had found two of its moons, and the largest moon of Neptune was discovered less than three weeks after the planet itself was first seen in 1851. Now and again, smaller and fainter moons were found, and by 1950 the tally of outer planet satellites ran: Jupiter, eleven; Saturn, nine; Uranus, five; and Neptune, two.

By this time such bodies had long been little more than astronomical curios. What was known about them other than the mechanics of their orbits (which were already known in great detail) could probably have been written in the space of this book so far. However, about this time people did begin to think seriously about what these bodies might be made of. Spectroscopic observations began to reveal a few curiosities about the composition of their surfaces, and, in the case of Titan (the largest moon of Saturn), about its atmosphere. Even so, these satellites remained specks of light to be observed using large telescopes. No one could say what any of them would look like close to; no one knew what forces had shaped their surfaces or what their internal structures were. All this changed in the dramatic ten and a half years between the encounters with the moons of Jupiter by the space probe *Voyager-1* in March 1979 and the fly-by of the Neptune system by *Voyager-2* in August 1989.

When geologists pay any attention to the geology of other worlds (which is usually not often enough) they think mostly of the other terrestrial planets: Mars, with its giant extinct volcanoes and dry canyons; Venus, with its plains, mountain belts, and greenhouse atmosphere; Mercury, with its impact-scarred surface bearing also the marks of a history of global compression; or our own Moon, with its impact craters and basalt lava plains. However, the *Voyager* missions showed other families of worlds, planet-sized in their dimensions, which have geological histories that are equally fascinating, if not more so (Fig. 1.1). Many of the moons of the outer planets became known as tangible worlds with a

Fig. 1.1. Part of the surface of Ganymede, a satellite of Jupiter, showing a surface scarred by a variety of of geological processes.

greater range of geological processes on display than almost anyone would have dared hope for. It is on these, the larger among the moons of Jupiter, Saturn, Uranus, and Neptune, that I will concentrate in this book. These are the worlds in their own right whose story I wish to tell.

1.1 Geology on planetary satellites

As will be discussed at greater length in subsequent chapters, it is possible to divide the satellites of the outer planets into several groups. Among the larger bodies—those with radii in excess of about 200 km—there is a continuum of types ranging from currently active worlds with continually erupting volcanoes, through worlds that have been volcanically and tectonically active recently (geologically speaking, at least), to those that have been inactive for so long that all traces of geological processes have been obliterated by impact craters. Some basic facts about these worlds are presented in Table 1.1, and the complete satellite families of the outer planets are shown to scale in Fig. 1.2. Two of these satellites are bigger than the planet Mercury (although being of lower density their masses are less), these and two others are both bigger and more massive than the Earth's Moon, and a total of five are bigger and more massive than the planet Pluto.

As Fig. 1.2 shows, in addition to the major satellites there is a plethora of smaller bodies, ranging in size down to the smallest detectable during the *Voyager* fly-bys (about 20 km across), which are generally irregular in shape and represent collisional fragments of larger moons or captured asteroids and comet nuclei (Fig. 1.3). While these would make a fascinating study, we know comparatively little about most of them. More importantly, from the point of view of this book, they are not really 'worlds' in that they never had histories that can be described in terms of geological processes. These will be largely passed over in what follows in order to concentrate on the larger moons.

Most of the larger moons are composed mainly of ice with variably rocky interiors, but two of them (Io and Europa) have silicate rock reaching outwards to, or almost to, their surfaces. With the exception of Titan, atmospheres are tenuous or absent, so the surface is unprotected from cosmic radiation, which may be made more severe in its effects by the influence of the planet's strong magnetic field. In general, the geological story of each of these worlds is written on a surface consisting essentially of ice. This does not mean that the outer solar system is a playground for glaciologists. The lucky people are geologists in general, because at the low temperatures prevailing so far from the Sun the ice behaves mechanically very much like silicate rock, both at the surface and at depth. Furthermore, in the outer reaches of the solar system the ice is not usually pure water-ice (H_2O); it is contaminated by such things as ammonia (NH_3) and methane (CH_4), which cause it to melt in ways analogous to the partial melting of terrestrial rocks and give rise to the possibility of a comparable range of volcanic and intrusive 'igneous' processes.

Some of the highlights of these worlds are introduced in the next few pages. Chapter 2 then takes a look at the origin and evolution of planetary satellites and reviews what we knew about them before the *Voyager* missions. Chapter 3 gives an overview of the

Table 1.1 The satellites of the outer planets, showing the characteristics of all those bodies whose radius exceeds about 200 km. Pluto is included because of its resemblance to Triton. Data for the Earth and Moon and the other terrestrial planets are included at the foot of the table for comparison. (For satellites with elliptical orbits, the orbital radius quoted is the semi-major axis.)

Planet	Satellite	Radius (km)	Mass (10^{20} kg)	Density (10^3 kg m^{-3})	Orbital radius (10^3 km)
Jupiter	Io	1815 ± 5	894 ± 2	3.57	421.6
	Europa	1569 ± 10	480 ± 2	2.97	670.9
	Ganymede	2631 ± 10	1482.3 ± 0.5	1.94	1070
	Callisto	2400 ± 10	1076.6 ± 0.5	1.86	1883
	12 others	<135			
Saturn	Mimas	197 ± 3	0.38 ± 0.01	1.17	185.52
	Enceladus	251 ± 5	0.8 ± 0.3	1.24	238.02
	Tethys	524 ± 5	7.6 ± 0.9	1.26	294.66
	Dione	559 ± 5	10.5 ± 0.3	1.44	377.40
	Rhea	764 ± 4	24.9 ± 1.5	1.33	527.04
	Titan	2575 ± 2	1345.7 ± 0.3	1.88	1221.85
	Iapetus	718 ± 8	18.8 ± 1.2	1.21	3561.3
	11 others	<175			
Uranus	Miranda	236 ± 3	0.75 ± 0.22	1.35 ± 0.39	129.8
	Ariel	579 ± 2	13.5 ± 2.4	1.66 ± 0.30	191.2
	Umbriel	586 ± 5	12.7 ± 2.4	1.51 ± 0.28	266.0
	Titania	790 ± 4	34.8 ± 1.8	1.68 ± 0.09	435.8
	Oberon	762 ± 4	29.2 ± 1.6	1.58 ± 0.10	582.6
	10 others	<85			<86
Neptune	Triton	1350 ± 5	$213.8 \pm$	2.075 ± 0.019	354.8
	7 others	<200			
Pluto	—	1142 ± 9	124 ± 30	2.07 ± 0.05	—
	Charon	596 ± 17	11 ± 3	2.07 ± 0.05	19.6
Mercury	—	2439	3300	5.4	—
Venus	—	6051	48700	5.3	—
Earth	—	6371	59970	5.517	—
	Moon	1738	734.9 ± 0.7	3.34	384.4
Mars	—	3394	6420	3.9	—

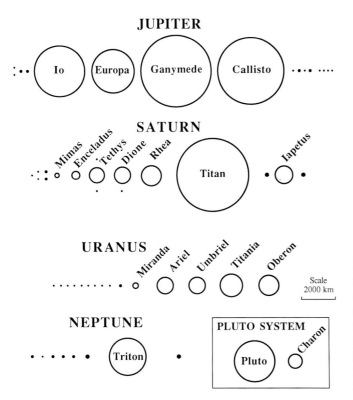

Fig. 1.2. The satellites of the outer planets, showing their correct relative sizes and ranged from left to right in order of distance from each planet. Names are shown only for those greater than about 200 km in radius, which are the main topic of this book.

remarkable *Voyager* project, including the scientific and political background as to how one spacecraft was able to visit four planets in turn. In Chapter 4 the ground rules are presented for understanding the geology of the larger moons in terms of their internal structure; they are layered bodies, and in particular the concept of a rigid icy outer layer (the lithosphere) overlying a weaker icy layer (the asthenosphere) is set up by analogy with the Earth's lithosphere and asthenosphere, which are made of silicate rock. The surfaces of the moons themselves are discussed in detail in the succeeding chapters.

1.2 A gallery of worlds

Those moons of the outer planets that have not been extensively resurfaced by geological processes are characterized by vast numbers of impact craters. These craters may occur in such profusion that they completely cover the surface, leaving no trace of its former state, as shown, for example, in Fig. 1.4. Most of the highlands of the Earth's own Moon are densely cratered in this way, but younger terrains have fewer craters, and variations in the numbers and sizes of craters from place to place are used to estimate the ages of the various units exposed on the surface.

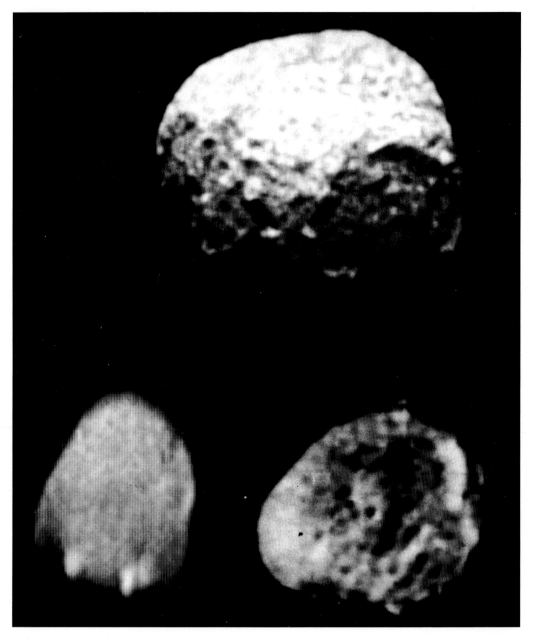

Fig. 1.3. The best views of some of the larger irregularly shaped satellites. These are perhaps fragments resulting from the collision of larger bodies. Moons such as this are too small for geological processes and are not considered in any detail in this book. *Top*: Proteus, a satellite of Neptune discovered in 1989 by *Voyager-2* (almost 200 km in radius). *Lower left*: Amalthea, an inner satellite of Jupiter (semi-major axes $135 \times 82 \times 75$ km). *Lower right*: Hyperion, an outer satellite of Saturn ($175 \times 120 \times 100$ km).

Fig. 1.4. Part of the surface of Rhea, a satellite of Saturn whose surface is saturated by impact craters. This view shows an area about 300 km across.

Such dating using cratering statistics works well in the inner solar system, as far as can be told with the limited independent (radiometric) dates from the Moon and none from the other planets. The standard model is that the oldest surfaces seen, those virtually saturated with craters, date from about four billion years ago. This was the time of the so-called late heavy bombardment by fragments left over from the formation of the planets 500 million years previously. The supply of impacting bodies began to decline rapidly after about 3.9 billion years ago and has remained at a more or less uniform rate during the past 3.8 billion years. Thus for surfaces younger than 3.8 billion years there is a fairly straightforward correlation between age and density of craters. As you will see in Chapter 5, in the outer solar system the calibration between the age of a surface and its crater-density is less certain, because there are alternative populations of impacting bodies available and heavy cratering may have occurred much more recently than it did in the neighbourhood of the Earth. Nevertheless, the older a surface is the more craters it will tend to have, so that a surface which is densely cratered cannot have been subject to resurfacing by geological processes such as volcanism or tectonism (faulting and other deformation) to any degree since the rate of cratering declined.

To the geologist then, the heavily cratered moons are the least interesting ones. However, even these have a tale or two to tell. For example, the cross-sectional shape of the craters, particularly the big craters, can tell us about the strength of the outer layers of the moon. If the topography of a crater is subdued and appears to have become flattened by subsidence, then it can be assumed that the outer layers are (or were once) thin or weak enough to deform under their own weight. This information can be used to constrain models of the moon's internal structure and thermal history. Chapter 5 will show that such properties are far from uniform even between moons that appear, at first sight, to look much the same.

On many of the other moons, terrains that preserve evidence of a period of dense cratering have been wholly or partly covered over or destroyed by younger surfaces formed either in response to faulting or as a consequence of resurfacing by volcanic processes (Figs 1.5 and 1.6). The height and steepness of fault scarps and other slopes can be used, just like craters, to give information on the strength of the outer layers of the moon. As will be discussed in Chapters 4, 6, and 7, the volcanism that has occurred is generally the eruption of ice-dominated flows, but in ways closely analogous to many terrestrial volcanic processes.

Whether the dominant resurfacing agency has been tectonic or volcanic, moons that show signs of such geological activity have evidently had enough internal energy at some time to drive these processes. This begs the question of what was the heat source. The small amount of rock in these bodies makes radiogenic heat (from the decay of radioactive elements), which drives the Earth's heat engine, an unlikely candidate except during the extreme youth of the solar system. Therefore other sources have to be considered, such as energy derived from gravitational interactions through tidal processes. The amount of energy required to cause the observed amount of resurfacing depends on how the composition, strength, and temperature of the body vary with depth, and so the observations that can be made of their surfaces have a great deal to tell us about how each of these moons has evolved. We shall return to this theme in Chapters 6 and 7.

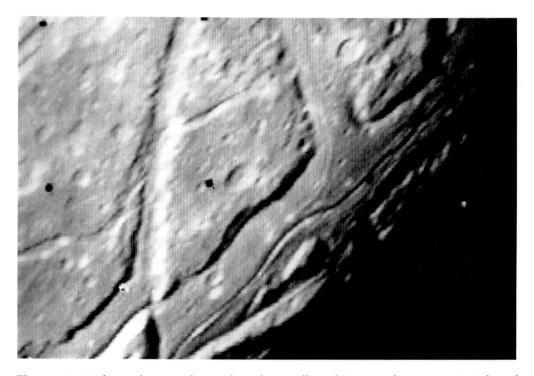

Fig. 1.5. A 400 km wide view of part of Ariel, a satellite of Uranus, showing ancient, heavily cratered terrain cut by a major fault-bounded valley. The valley floor has been filled by a volcanic flow unit that is evidently much younger than the surrounding terrain because of the comparative sparseness of craters in this area.

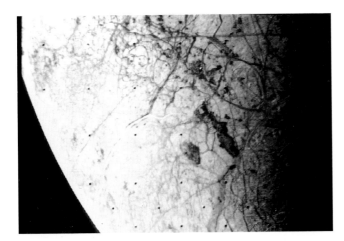

Fig. 1.6. A 2000 km wide view of part of Europa, a satellite of Jupiter. The surface is cut by a variety of tectonic features, and because there are very few impact craters it is evidently very young.

Fig. 1.7. The image on which the first volcanic plumes on Io, a satellite of Jupiter, were discovered. The bright spot to the left of the crescent disc is a plume rising into the sunlight from the Loki fissure, and the top of a fountain-like second plume is faintly visible rising some 300 km above the bright limb of Io, towards the lower right, from the Pele eruptive centre.

The jewels in the crown of solar system exploration have been those worlds where geological processes have actually been observed in action. The most remarkable among these are Io and Triton. Io was one of the first moons to be imaged by *Voyager-1*. Blending geophysical prowess and impeccable timing in equal measure, Stanton Peale and

Fig. 1.8. General view of Triton, a satellite of Neptune, showing a variety of terrains. The south polar cap, overlain by dark streaks representing dust deposits from wind-blown eruption plumes, occupies the left of the image. A colour view of the same region is shown in Plate 7.

colleagues proposed a model predicting that the interior of Io must be largely molten because of tidally generated heat, published in the journal *Science* just a few days before *Voyager-1*'s encounter with Io. Their theory was confirmed dramatically by *Voyager* images showing not only a surface dominated by fresh-looking volcanic features that obliterate all traces of impact craters, but also nine active volcanoes with eruption plumes reaching over 300 km in height (Fig. 1.7).

Triton was the last object in the solar system to be visited by *Voyager-2*. Some people had hoped to see oceans of liquid nitrogen there. They were frustrated in that respect, but the surface contains evidence of such a wide variety of tectonic, volcanic, and other geological processes that Triton was far from being a disappointment. Triton has arguably the most fascinating, and least-well understood, surface of all the outer planet satellites (Fig. 1.8). In addition to this the images showed several geyser-like plumes that reach an altitude of 8 km before being driven down-wind in the tenuous atmosphere.

Io, Triton, and two other probably active moons will be discussed at length in Chapter 7. Chapter 8 presents what we know about the geology of Titan, the second largest satellite in the solar system (which is not very much because Titan's atmosphere hides its surface), and then the planet Pluto (which may be similar to Triton) and its satellite Charon. For now however, you are invited to come down to Earth and consider the pre-*Voyager* view of planetary satellites and their significance in understanding the formation and evolution of the solar system.

2

Composition and evolution of satellites

And now a bubble burst, and now a world
Pope, *An Essay on Man*

When Jupiter is at its closest to the Earth (just over 600 million km), the largest of its satellites appears less than two seconds of arc across. This is about as big as your thumbnail seen from a distance of three kilometres. Satellites of the other outer planets are even further away and physically smaller, so it is not surprising that Earth-based observations have been able to provide so little information about their geology.

Thus, in a 550 page astronomical text written around the turn of the century, the author has this to say about the surfaces of the galilean satellites:

Owing to the minuteness of the satellites as seen from the earth, it is extremely difficult to perceive any markings on their surfaces, but the few observations made seem to indicate that the satellites (like our moon) always turn the same face towards their primary. Professor Barnard has, with the great Lick refractor, seen a white equatorial belt on the first satellite, while its poles were very dark. Mr Douglass, observing with Mr Lowell's great refractor, has also reported certain streaky markings on the third satellite.*

The Lick refractor had a 90 cm aperture, and Lowell's instrument (with which he made his famous, but illusory, observations of the canals of Mars) had an aperture of 61 cm; these are among the biggest and optically most precise conventional refracting telescopes ever made, and the nature of the observations that were made with them shows the limitations of ground-based observations in mapping satellite surfaces. Not surprisingly, the same author has nothing at all to say about the surfaces of Saturn's satellites.

So far as they go, the descriptions just quoted of the pale equatorial region on Io (the first satellite) and the general streaky nature of Ganymede (the third satellite) are now known to be broadly correct, but they did not cast much light on what the surfaces of these worlds are really like. However, the realization that these bodies are in synchronous rotation, so that they rotate on their spin-axis once during the course of each orbit, was an important discovery. In the next section we will discuss what this and other observations that can be made from the ground tell us about the nature of planetary satellites.

---

* Sir Robert S. Ball, *The Story of the Heavens*, Cassell & Co., first published 1886.

2.1 Planetary satellites seen from Earth

We now know that all the outer planet satellites greater than about 200 km in radius are in synchronous rotation. This can hardly be coincidence. What it implies is that the gravitational, or tidal, influence of the planet around which each one orbits has slowed the satellite's original rotation down until it matches the orbital period. The process is understood to have operated as follows. The gravitational attraction of the planet distorts the shape of the satellite by raising a tidal bulge on it. Ideally this bulge should always face directly towards the planet, but in a real material there is no way in which such a bulge can keep pace if the satellite is spinning at all rapidly. The bulge therefore lags behind the direct alignment, just as high tides in the Earth's oceans are out of phase with the position of the Moon in the sky. The gravitational force exerted by the planet on this out of line bulge eventually drags the bulge into line, which it can do only by slowing down the satellite's rotation until it matches the period of the orbit. Once this has been achieved the bulge has become stationary on the surface of the satellite, and, provided that the satellite is in a circular orbit, no further work is required to keep the bulge in place.

All this is presumed to have happened early in each satellite's history. The calculated time-scales for bringing most satellites into synchronous rotation are in the range 10^4–10^7 years. A crucial consequence of this tidal 'despinning' is that it must have generated a large amount of heat within the satellite. As will be seen, the magnitude and duration of such heating events have important consequences for the development of internal layering and the operation of tectonic and volcanic processes at the surface.

Telescopic observations also enabled the masses of several of the satellites to be measured from calculations based on the perturbations they cause to each others orbits; so the extremely detailed orbital data referred to in Chapter 1 came in useful after all! The masses that were derived for Io, Europa, and Ganymede by this method in 1921 are within five per cent of their currently accepted values. However, to work out what the satellites are likely to be made of it is important to know their densities. The only ground-based way to determine density is to calculate it from mass and volume. The angular size of a planetary satellite seen from Earth is so small that it is difficult to measure precisely, and because volume depends on the cube of the radius, the errors in volume, and hence in density, determined by ground-based observations were several times greater than those for mass. Nevertheless, it was possible to demonstrate that the inner galilean satellites (Io and Europa) are denser than the outer two (Ganymede and Callisto). This is a striking parallel with the solar system itself where the inner planets are denser than the outer ones; a theme to which we shall return later.

Even knowing the density of a satellite does not tell us what it is made of. A second important line of evidence that can be gathered from ground-based observations—and this is the one that enables us to rule out the proverbial green cheese and other fantastic materials—is the analysis of the light reflected by these bodies using spectroscopic techniques. When matter reflects light the molecules within it absorb certain wavelengths, resulting in a series of absorption bands in the reflected spectrum that are characteristic of the composition of the surface. Thus, by measuring the way that the intensity of sunlight reflected by a satellite varies with wavelength, it is possible to understand the composition of the surface material. To make sure that only the absorptions due to substances on the satellite itself are measured, it is necessary to make a correction for the shape of the Sun's spectrum and also to allow for absorption by gases in the Earth's atmosphere above the observatory.

The first breakthrough with this technique came in 1944 when Gerard Kuiper found absorption features due to methane in the reflectance spectrum of Titan. As Kuiper realized, the particular features he saw could be due only to methane in the gaseous state, which proved that Titan (alone among the planetary satellites, as is now known) has a substantial atmosphere. In fact, the atmosphere of Titan is so dense that it hides the surface completely, and its nature is still not known (Chapter 8). There is no such problem with the other satellites, and from the late 1950s onwards Kuiper and his successors were able to demonstrate the presence of water-ice dominating the visible and near infra-red reflectance spectra of many of the other major satellites (Fig. 2.1). A notable exception is Io where there was evidently little or no water; instead Io has a highly reflective, markedly red surface, which, during the 1970s, some workers began to attribute to the presence of sulphur.

The detection of ice on most of the planetary satellites was certainly compatible with the very low surface temperatures (120–170 K) of the galilean satellites first revealed during the 1960s by the use of newly developed thermal infra-red telescope instruments. However, the trouble with relying on spectral information is that the compound that dominates the spectrum is not necessarily the most abundant one on the surface; in addition, the technique only gives information about, at most, the outer few millimetres of the satellite. What if it has a thin coating that is not typical of the body as a whole? The thermal infra-red determination of the rate at which satellites cool as they pass into the shadow of their planet shows that they generally have a low thermal conductivity, so that at

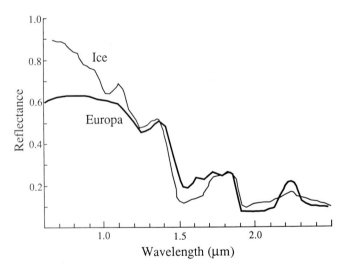

Fig. 2.1. The reflectance spectrum of Europa compared with that of coarse-grained water-ice, measured between 0.6 and 2.5 μm wavelength. The visible spectrum extends only as far as 0.7 μm, so this figure shows mostly the reflected infra-red part of the spectrum.

least the outermost metre or so is likely to consist of loose fractured and broken material, rather like the lunar regolith of impact-generated debris in which the *Apollo* astronauts left their footprints. If the regolith on the satellites is the result of an impact process (as the cratering we have since observed on most of them indicates), then this suggests that there must be a considerable degree of homogenization of the outermost material to some depth. Unfortunately, this does not rule out the presence of frosts and other ephemeral deposits that could still dominate the spectrum.

Despite the reservations quoted above, by the early 1970s it had become fairly clear that, with the exception of Io and Europa, the satellites of the outer planets for which there were reasonable data have fairly low densities, in the range 1.2–2.0×10^3 kg m^{-3}. This is less than twice the density of water, and only about half the density of the rock that makes up the Earth's crust. A reasonable assumption at the time, which is still regarded as good, is that these bodies are composed of a mixture of rock and ice. The domination of their reflectance spectra by ice combined with the fact that their density is generally greater than that of ice is a good indicator that their exteriors are icy and their interiors rocky.

This concept was (and remains) compatible with models of how the satellites were assembled, and the nature of the material from which they could have formed early in the life of the solar system. It is worth pausing briefly, before following the story of the exploration of the planetary satellites any further, to review what we can infer about their origin and early evolution.

2.2 Origin of planetary satellites

This section gives an overview of the origin and evolution of planetary satellites. It is a post-*Voyager* view, although in many of its essentials it resembles models that were developed during the 1970s. To understand how the planetary satellites were formed, we must first consider how the solar system as a whole developed.

2.2.1 Origin of the solar system

It is almost universally accepted that the Sun and its planets grew from a cloud of gas and dust. As this cloud contracted under its own gravity, much of the mass became concentrated at the centre, in the proto-Sun. The cloud originally had some degree of rotation, so, as the centre contracted, the conservation of angular momentum forced the rest of the cloud into a disc, which was flattened and rotating in the same plane as the Sun itself. This disc is usually called the solar nebula. It probably took not much more than 10^4 years for this process to occur. By the end of that time nucleosynthesis had started within the Sun and it was shining hotly. Various very young stars in this stage of evolution, surrounded by dense clouds of gas and dust, have been recognized over the past decade, particularly with the aid of infra-red astronomy.

The inner parts of the solar nebula were hot, even before the Sun's nuclear reactions began, because the proto-Sun would have been radiating strongly as a result of gravitational energy lost during contraction. With the passage of time, the solar nebula became less dense as a consequence of two processes: the loss of material to interstellar space and condensation to form solid objects. This lower-density nebula allowed heat to be radiated away more easily, leading to a general cooling of what remained of the nebula. At some time during this cooling and dissipation stage, the material that now forms the planets condensed and aggregated from the nebular material.

So what was this solar nebula made of? Fortunately we can determine its initial composition quite easily, if we are prepared to accept that the outer layer of the Sun has remained an unadulterated sample of this material. As the Sun is so hot, its elemental abundances can be measured spectroscopically simply by measuring the depth and wavelength of the absorption bands in its spectrum. The solar abundances of the most common elements are listed in Table 2.1; primitive meteorites have similar elemental abundances, which greatly strengthens the argument that both they and the Sun are reliable samples of solar nebula material.

Condensation of much of the solar nebula material probably took place over an interval of about 10^4–10^6 years. As indicated in Table 2.2, the earliest material to condense was composed of the most refractory molecules, that is those having the highest condensation temperatures. These include elements that are not in the top ten of the abundance charts, such as the oxides and silicates of calcium, aluminium, and titanium. We have direct evidence of this in the form of tiny grains of these substances that are found as inclusions in many meteorites. As the temperature of the nebula decreased, nickel and iron condensed, followed by more familiar rocky minerals, such as magnesium-rich silicates like olivine and pyroxene. It is clear that much of this matter collected into globules from about a millimetre to a centimetre across, because these are preserved as 'chondrules' embedded

Table 2.1 Solar (cosmic) abundances for the ten most common elements, normalized (by convention) to Si = 1.0×10^6. These abundances are assumed to have been much the same in the solar nebula

Element	Abundance
H	28.2×10^9
He	1.8×10^9
O	16.6×10^6
C	10.0×10^6
N	2.4×10^6
Ne	2.1×10^6
Si	1.0×10^6
Mg	850×10^3
Fe	700×10^3
S	460×10^3

Table 2.2 The temperatures at which some notable constituents of the solar nebula may have begun to condense. Hydrogen, neon and helium could have condensed only below 30 K and their presence in the gas giant planets is almost certainly due to gravitational capture as gases; these elements are rare in smaller bodies, including planetary satellites

Temperature (K)	Mineral
1758	Corundum, Al_2O_3
1647	Perovskite, $CaTiO_3$
1513	Spinel, $MgAl_sO_4$
1471	Nickel–iron metal, Fe, Ni
1450	Diopside (pyroxene), $CaMgSi_2O_6$
1444	Forsterite (olivine) Mg_2SiO_4
700	Troilite, FeS
550–330	Hydrated minerals
180	Water, H_2O
120	Ammonia, as $NH_3 \cdot H_2O$
70	Methane, as $CH_4 \cdot 6H_2O$
70	Nitrogen, as $N_2 \cdot 6H_2O$

in primitive meteorites (Fig. 2.2). Radiometric dating shows that chondrules formed over an interval of no more than a few hundred thousand years at about 4.5 billion years ago, which is normally regarded as the age of the solar system. At lower temperatures, progressively less refractory (i.e. more volatile) constituents of the solar nebula condensed, such as iron sulphides. Of the more abundant elements in Table 2.1, only oxygen was a significant contributor to the condensates up to this point, but at lower

Fig. 2.2. Close-up of a fragment of the Allende meteorite. Roughly spherical chondrules, representing originally liquid droplets, are imbedded in a carbon-rich silicate matrix. The large pale chondrule above left of the centre is about 2.8 mm across.

temperatures it became possible to begin to mop up the hydrogen (in water, H_2O), carbon, and nitrogen (in ammonia, NH_3). Perhaps the greatest remaining uncertainty from the point of view of interpreting the composition of bodies in the outer solar system is the form in which most of the carbon resides. This depends on the temperature at which the atoms in the solar nebula equilibrated to form molecules, and on the degree of availability of oxygen at the time, with the result that most of the carbon could have been taken up either in methane (CH_4) or carbon monoxide (CO). To complicate the issue further, some models call for a proportion of elemental carbon (e.g. graphite) as a condensate, and it is clear from carbonaceous chondrite meteorites and studies of the nucleus of Halley's comet that much of the carbon ends up in organic compounds.

While this sequence of cooling and condensation was taking place, the condensed material was aggregating into progressively larger masses, at first because those particles whose random motion brought them into contact with each other tended to adhere, and eventually because the gravitational attraction of the larger of the planetesimals that formed in this way resulted in an increasing rate of collision and accretion on to them by smaller planetesimals.

Probably the most important unresolved question about the whole process of planetary accretion is the rate at which it occurred relative to the rate of cooling of the solar nebula. If the rate of cooling was slow, then the planets probably accreted heterogeneously, that is the more refractory material would have had time to condense into grains, aggregate into planetesimals and to accrete into protoplanets before the more volatile material became available. On the other hand, if the rate of cooling was relatively fast, then there could have been dust particles with a whole range of compositions coexisting at one time, so that the planets would have accreted in a homogeneous fashion. On the heterogeneous accretion model, the observed layering of the planets (dense, mainly refractory, interiors with less refractory outer layers, surrounded by a hydrosphere and atmosphere) is a natural consequence of sequential accretion. On the homogeneous accretion model, the layering is a result of later gravity-driven differentiation and the outwards migration of volatile material. Either scenario can be adapted to explain the generally decreasing densities of the inner planets (Mercury, Venus, Earth, and Mars) outwards from the Sun and their deficiency in light, gaseous elements compared with the outer giant planets, by stipulating that temperatures within the solar nebula in the inner solar system remained too high for the more volatile constituents to condense. For example, it is unlikely that water-ice could have condensed directly from the solar nebula any closer to the Sun than about five times the Earth's orbital radius (i.e. 5 astronomical units, or AU).

These homogeneous and heterogeneous accretion models represent extreme variants of what could have taken place. It is generally accepted that the truth lies somewhere between the two. What then does this tell us about the satellites of the outer planets? Treated at its simplest, it suggests that, if they formed roughly where they are now seen, they should contain progressively more volatile material as the distance from the Sun increases. As you will see in Chapters 5–7, this is compatible with the evidence that the water-ice is contaminated with volatiles such as ammonia by the time the Uranus system is reached, with the addition of methane and nitrogen in the Neptune system. The first appearance of ice-dominated bodies at Jupiter is compatible with the 5 AU inner limit for the condensation of ice, as Jupiter's orbit has a radius of 5.2 AU.

2.2.2 *Satellite formation*

But how did the satellites form? Was their creation part and parcel of the formation of their planet, or did they form independently and become captured later? One of the most important pieces of evidence in this respect that we have not yet touched on is that all the major planetary satellites except Triton orbit their planet in the same direction as the spin of the planet. The inclination of their orbit to the plane of their planet's equator is very small, usually less than 1°. Capture of all these moons into orbits of this nature is unlikely, and for Jupiter, Saturn, and (perhaps) Uranus, the major satellites are generally regarded as having been formed by condensation from a disc of gas and dust around each primitive planet, each rather like a smaller version of the original solar nebula. It is not clear whether such a disc was formed by material that was shed by the protoplanet as it shrank, or whether it grew around it, either by scavenging gas and dust from the solar nebula, or by collecting debris from collisions between planetesimals. Either way, the temperature would have increased towards the centre of such a disc, providing an opportunity for the chemical composition of the satellite-forming material to evolve.

Whatever the origin of such a protosatellite disc about a planet, it must have developed into a satellite system in much the same way. Small particles would have aggregated into larger ones that would, over time, have collided. Eventually the largest bodies would have swept up all the debris in their vicinity to form the moons we see today. The amount of heat, and consequent melting, liberated in this accretion process must have been an important factor in controlling the thermal evolution and differentiation of the satellite, and this is an issue to which we shall return in Chapter 4. The outwards decreasing density of the galilean satellites of Jupiter is usually taken to imply that there was less capture or retention of volatiles closer to the proto-Jupiter, as a result of elevated temperatures in this region caused by thermal radiation from the still-contracting planet. As was remarked earlier, in this respect the Jupiter system mimics the solar system in miniature. However, no such progression in densities is evident among the moons of Saturn or Uranus (Table 1.1), and it has recently become apparent that some or all of these satellites may have long and complex histories, so that they no longer retain their original size or mass distribution.

There are various exceptions to this model of the origin of satellites. The first of these are most of the small satellites of the outer planets. Some of these are icy bodies with near-circular orbits, which may represent fragments from collisions involving larger moons. Others that appear to be rocky or carbonaceous in composition, and those that are icy but are in strange orbits, may equally well be captured asteroids or comet nuclei. The second prominent exception is Triton, the only really large moon of Neptune. This has a retrograde orbit, that is, its orbital motion is in the opposite direction to the planetary spin. Moreover, the orbit is inclined to the planet's equator at a considerable angle (21°). The most probable explanation is that Triton was formed elsewhere in the outer solar system but was captured by Neptune's gravity after a close encounter with the planet. This idea is supported by the fact that, as far as we can tell, the planet Pluto is more like Triton than anything else in the solar system. There is no satisfactory explanation as to how Triton's capture may have occurred, but as you will see in Chapter 7 this event may also have been the cause of some of the strange geological events that have taken place there.

The Uranus system poses similar problems. The planet has five major satellites (Table 1.1), all of which are in near-circular prograde orbits in the plane of the planet's equator. So far, so good: it looks not unlike Saturn's family of moons. However, the whole system is tilted on end, so that the axis of the planet's rotation lies at an angle of 98° to the plane of its orbit (another way of looking at this is to say that it rotates in a retrograde sense, on an axis inclined at 82°). For comparison the axes of the three other gas giants are inclined at less than 30° to their orbits, and the Earth's axis is tilted at a little over 23°.

It is likely that the rotation axis of Uranus originally had a similar orientation to that of the other planets, because it should have inherited its rotation from the nebula out of which it condensed. One way for Uranus to have been knocked over at some stage after its formation is if a large body (bigger, for example, than any of its present satellites) collided with it. Such an impact could have created satellite-forming debris considerably later in the planet's history than most other models require. Interestingly, there are geochemical grounds for supposing that just such a giant impact was responsible for the origin of the Moon, by condensation of the material knocked out of the mantle of the primitive Earth, but the grounds for demonstrating a similar process at Uranus are much weaker. However, it is worth noting that if the moons of Uranus existed before the planet's tilt was changed, this re-orientation, however it happened, would have been almost certain to have badly disrupted them. Tidal processes can be called upon to drag the satellites' orbits back into line with the new rotation axis, but only over a prolonged period during which it is likely that there would have been several collisions between satellites or between satellites and large fragments from disrupted satellites. The resulting debris would have led to a new interval of satellite accretion.

The matter of what happens to an icy moon after it accretes, and in particular how it heats up internally and where and when melting may occur, was the subject of some authoritative and thought-provoking studies during the final pre-*Voyager* years. These, and some developments arising from them, are discussed in the next section.

2.3 Heating and differentiation of an icy moon

To understand this topic requires a simple appreciation of ice physics and planetary heat sources. Let us assume for now that a planetary satellite accretes homogeneously, that is, it grows by the collision of lumps composed of an ice–rock mixture. The densities of most of the satellites (Io and Europa excepted) indicate that the mixture was fairly close to the cosmic abundance ratio of ice to rock, which is about 60:40 by mass. In this chapter we will usually keep things simple by regarding the ice as being just water-ice, but you should be aware that adding ammonia or methane to the ice will slightly decrease its density and, in the case of ammonia in particular, have important consequences in lowering the melting temperature and also making it easier for the ice to convect in the solid state. These matters will be discussed further in Chapter 4.

An important consideration is that when the volatile compounds listed in Table 2.2 condense in a near-vacuum, they do so at temperatures considerably below their normal melting point, and go directly from vapour to solid. This means that in a homogeneously accreted satellite, in the absence of a heat source, there would be no melting and hence no way in which the rock could separate from the ice. The body would therefore retain its initial mixed, or undifferentiated, structure.

2.3.1 *Accretional heating*

Perhaps the most obvious way in which a satellite forming in this way can warm up is by the kinetic energy liberated by the impact of the accreting bodies. How much of this energy is retained within the satellite and how much is immediately radiated away to space is debatable; probably only about 10–50 per cent is retained. Several (but not all) models put forward during the past twenty years or so have tended to agree that, because they are so massive, Ganymede and Callisto had probably been warmed sufficiently by accretional heating to have melted by the time their radius had grown to more than about 1000–2000 km. This would have freed the rock particles and fragments embedded within the ice, which would then have sunk and accumulated at the centre of the body. The result would be a differentiated, layered, structure with a rocky core surrounded by a rock-free water layer. Incidentally, the sinking of the heavier fragments would itself contribute about a further 10 per cent to the heating, through the loss of gravitational potential energy. Once the input of accretional heat had ceased, the water would freeze from the outside inwards, as heat was radiated to space. Note that the basic layered structure arrived at in this way is the same as that produced by heterogeneous accretion, in which the rock would have accreted first and the ice later.

Similar models for smaller bodies such as the moons of Saturn (Titan excepted) show fairly convincingly that these would never have collected or retained enough accretional heat for melting to occur. To predict the thermal evolution of these worlds, and to model the post-differentiation history of larger bodies, other heat sources need to be considered. The Earth is hot inside primarily as a result of the decay of radioactive elements, and this mechanism is another way in which the planetary satellites could have been heated.

2.3.2 *Radiogenic heating*

Evidence from some meteorites suggests that they were heated very soon after they formed, by the energy released from the decay of ^{26}Al, a short-lived isotope of aluminium with a half-life of just 7×10^5 years. The only known way for ^{26}Al to form is by nucleosynthesis in a star, and it is presumed to have been dispersed into space by a supernova explosion or similar catastrophe shortly before the formation of the solar system. As a result of its short half-life, all but the minutest trace of ^{26}Al has long since disappeared from the solar system, although decay products from ^{26}Al are found in many meteorites in amounts that demonstrate that these meteorites contained ^{26}Al when they were created. This means that the solar nebula must have formed and condensed within just a few half-lives of the event that distributed the ^{26}Al. However, most workers suppose that the formation of the planetary satellites, from meteoritic and other debris, occurred too long after the event for enough ^{26}Al to have survived for it to have played a significant role in even their initial heating.

The other radioactive elements in solar system rocks have much longer half-lives. The major source of post-accretionary heat in the Earth, the Moon, and the other terrestrial planets is from the decay of isotopes of uranium (^{235}U and ^{238}U), thorium (^{232}Th), and potassium (^{40}K). The half-lives of these isotopes are of the same order of magnitude as the present age of the solar system. The rate of radiogenic heat production from this source must therefore have been significantly greater when the solar system was young. Models for the rate of radiogenic heat production in the rocky fraction of icy satellites usually assume that it has similar concentrations of these elements to the concentrations we can measure on Earth, the Moon, and in chondritic meteorites. Taking account of the rate at which such heat would have been dissipated, the models generally show that moons less than about 600 km in radius are unlikely ever to have become warm enough for internal melting to have occurred. This means that if these satellites accreted homogeneously they must have remained undifferentiated unless an additional heat source operated. The same models suggest that it would have been possible for the ice in the interior of a larger moon to melt, because of the decline in radiogenic heat production over time, the layer of liquid water would subsequently have begun to freeze at its upper boundary.

The consequences of one such model, dating from 1978, for a moon of 700 km radius are shown in Fig. 2.3. For a body this small, the accretional heating is negligible, so at the time of formation the temperature is shown as uniform throughout (t_0 in Fig. 2.3). Radiogenic heat from within the satellite is unable to escape faster than it is generated, so the interior warms up. It begins to melt after about 0.6 billion years, and reaches a maximum temperature at 2.0 billion years (t_1 in Fig. 2.3). At this stage a core of rock has formed, by the sinking of the rock within the melt. The layer above is liquid water, which is kept at a uniform temperature (just above the freezing point) as a result of convection. Above the water is a layer that has not melted, and which therefore retains its initial composition. From this time onwards there is cooling throughout as a result of declining radiogenic heat production, which is now exceeded by the heat loss to space. This results in the present temperature profile (t_2 on Fig. 2.3), and in this example no liquid water layer remains. The internal structure has evolved through time as follows: initially an un-differentiated mixture of rock and ice (t_0); at the time of maximum melting (t_1) at 140 km radius core of rock, mantled by a 180 km layer of liquid water, overlain by a 380 km layer of undifferentiated rock and ice; at the present day (t_2) the same layered structure as the previous stage, except that the water layer has frozen right through.

This sort of model makes it reasonable to envisage the melting of a medium-sized satellite at some time in the past, which would have led to an at least partially differentiated structure, even if the accretion had been homogeneous. In larger moons, the melting

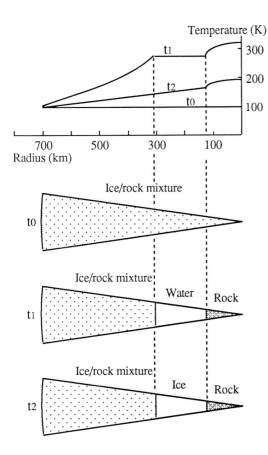

Fig. 2.3. (*Top*) Thermal profiles through a radiogenically heated 700 km radius moon (60% water-ice, 40% rock). t_0, immediately after formation; t_1, after 2.0×10^9 years; t_2, after 4.5×10^9 years. (*Bottom*) Internal composition of the moon at the same times.

would have reached closer to the surface before the radiogenic heating was outpaced by heat loss to space. However, any liquid water remaining would by now lie below a very great thickness of ice.

We have already remarked on a third heat source for some of the planetary satellites earlier in this chapter, tidal heating, and we must return to this now to complete our survey.

2.3.3 Tidal heating

As described in Section 2.1, until the rotation of a satellite becomes synchronous with its orbital period, heat is generated internally. This is a result of the continual flexing of the moon as a whole that is necessary to allow a tidal bulge to propagate round the globe. This could have been an important source of additional heat early in the life of a satellite, although its magnitude is not well constrained, or in the interval following capture in the atypical case of Triton. More generally, however, there is a significant source of tidally generated heat that continues to operate even today for some synchronously rotating satellites.

The rotation of a satellite on its axis is controlled by the conservation of angular momentum, and so the rate of rotation is effectively constant. Similarly, the speed at which a satellite orbits its planet is described by Kepler's laws of planetary motion, which state that the speed is greatest when the satellite is closest to its planet. For a perfectly circular orbit, the orbital speed is uniform and will keep exactly in pace with the rotation of a synchronously rotating satellite. This means that any tidal bulge always points exactly towards the planet: no work needs to be done to try to keep it in line, and no tidal heat is generated. However, for a synchronously rotating satellite in an elliptical orbit, the changes in orbital speed mean that when the satellite is closest to the planet (and so moving the fastest), its rotation lags slightly behind exact synchroneity, and when the satellite is furthest away its rotation gets slightly ahead.

The result of this is a slight wobble, or libration, in the satellite's orientation as seen from the planet. Work is then expended to try to move the tidal bulge into step with this changing orientation. This causes internal heating. In the absence of external influences, the eventual outcome would be gradually to nudge the satellite's orbit towards an exactly circular shape. However, this is not found in reality. The orbits of all the major satellites of the outer planets have measurable eccentricities, although these are less than about 0.02 in all cases (which means that when drawn on paper they are indistinguishable from circles). The reason for the persistence of orbital eccentricity is that the gravitational attraction between one satellite and another affects the shape of its orbit. The effect is particularly pronounced for the inner three galilean satellites of Jupiter because of the phenomenon of orbital resonance. For every one orbit made by Ganymede, Europa completes two orbits and Io four; this periodic alignment is enough to force their orbital eccentricities to remain far higher than they would otherwise be. The result is that their tidal bulges are continually flexed due to their changing orientation with respect to Jupiter's gravitational field. The amount of heat actually generated depends on the thickness and elasticity of layers within each satellite, for which estimates vary. In the case of Io, heat is generated by tidal dissipation at probably about 40 to 100 times the rate of radiogenic heating. This is enough to melt at least some of Io's interior, and, as noted in Chapter 1, a prediction based on such a model proved spectacularly successful when *Voyager-1* discovered Io's active volcanoes. A similar kind of extra heating of certain icy satellites may have been important in the past, especially as such bodies require less heat input before they melt.

2.3.4 *Implications of heating and differentiation*

Thus there are three main mechanisms by which the interior of a satellite may be heated: accretional, radiogenic, and tidal heating. In reality, these will all have contributed to the thermal history of the body, although perhaps not sufficiently to have had any noticeable effect. In general, it is expected that the biggest satellites must have experienced more accretional and radiogenic heating than the smaller ones. The strength of tidal heating depends on the history of orbital resonances with the adjacent satellites.

One of the most important implications of the models discussed is that melting and the consequent differentiation may have affected only the interior of a moon, leaving an outer layer with the original primordial rock and ice mixture. Rapid cooling by radiation to space may even make it possible to retain an unmelted outer layer on bodies big enough to have melted throughout by accretional heating. In general though, it is unlikely that the smaller major satellites ever experienced substantial melting, so if they accreted homogeneously they may still be largely undifferentiated.

There are various complicating factors and unknown parameters that make generalizations dangerous to use as guides for understanding the evolution of a planetary satellite. The factor that is perhaps the simplest to appreciate is that a mixture of rock and ice is denser than water, and so the outer layer of a partly differentiated body such as that in Fig. 2.3 at t_1 is gravitationally unstable. If the liquid water shell reaches close enough to the surface, the whole outer layer could fracture and founder, displacing the water towards the surface and leading to a reversal of the outer two layers. The result would be pure ice on the outside and mixed rock and ice in the middle layer.

Another uncertainty in the modelling is in the properties of ice at the high pressures and low temperatures that prevail within a planetary satellite. In particular, how ice deforms at very low (but geologically reasonable) strain rates cannot be measured under laboratory conditions. Seismic data show that the Earth's mantle is solid, yet it is evident that convective forces cause it to flow at rates of the order of 1 cm yr^{-1}. In the simple model of radiogenic melting presented in Fig. 2.3 it was assumed that all the heat loss from the ice and ice–rock mixtures is by conduction. However, if the ice is able to flow at a rate comparable to the rate of flow of the Earth's mantle, then it can transport heat outwards at a rate comparable to the conductive heat flow. By removing internal heat in this way, solid-state convection by ice would inhibit or even prevent melting.

One of the major complicating factors is that the molecular arrangement of the crystalline structure of ice is dependent on temperature and pressure. The ice-cube with which you may be familiar in your favourite drink is of the form known as ice I. If you were gradually to compress it (keeping it cold at the same time) there would come a point when its structure switched to an alternative, more closely packed, arrangement of molecules. In fact, there are several different solid phases of ice, the stability fields of which are shown in Fig. 2.4.

This figure is what is known as a phase diagram. The lines on it are the boundaries of the pressure–temperature conditions at which the various forms, or phases, of ice are stable. Thus at a temperature of 200 K, ice I will change to ice II when the pressure is increased to about 1.6 kbar. The upper, curved line on the diagram, known as the liquidus, shows the conditions under which the ice will melt. Note that a sample of ice II cannot be melted directly. On warming it changes to another form of ice before it reaches the liquidus.

The depth represented by a given pressure is different for every moon, because their different masses affect the local gravity. The depth–pressure relationship also depends on the average density, which is controlled by the exact proportions of rock and ice present, and the density gradient, which is controlled by the degree of differentiation. Generally speaking, however, the pressure inside an icy moon smaller than about 600 km in radius will be low enough to allow ice I to be stable all the way to its centre. In moons the size of Iapetus and Rhea (700–800 km in diameter) the pressure near the centre is large enough

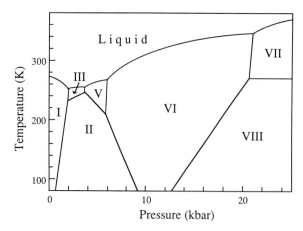

Fig. 2.4. Simplified phase diagram for pure water-ice.

to make ice II the stable phase, whereas the pressure–temperature gradient in a large moon such as Callisto is likely to take it in turn through the stability fields of ice I, ice II, ice VI, and then ice VIII, moving inwards. Figure 2.5 shows two extreme models for Callisto, indicating the zones of stability of its ice phases.

The fact that ice has several phases that are stable under different conditions influences the evolution of an icy moon in two ways. In the first place, it is difficult for solid-state convection to occur across phase boundaries, so convective heat loss in the larger moons that have several internal phase transitions will tend to be inhibited. The second involves consideration of what happens to the solid part of a moon as it cools. In particular, a phase change from ice VI to ice VIII would occur within the larger satellites simply as a result of falling temperature. For example, at a pressure of 18 kbar ice VI inverts to ice VIII as soon as the temperature drops below 200 K. This phase change is associated with a decrease in volume, and would be expected to result in traces of compressional tectonics at the surface.

A final complication is the influence of volatile contaminants within the ice, such as ammonia and methane, especially at Saturn and beyond where these should have been able to condense from the solar nebula. Both these can mix with water-ice but their abundances are debatable and how their presence influences the strength of ice, and hence its ability to

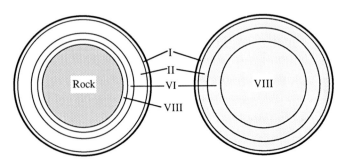

Fig. 2.5. Fully differentiated (*left*) and undifferentiated (*right*) models for Callisto. The former has a silicate core (dense stipple) surrounded by pure ice, the latter has a mixture of rock and ice throughout. The ice phases that are stable in each zone are indicated by roman numerals.

convect, is not yet well understood. Contaminants also lower the melting point of the ice; in particular, if there is any ammonia present in the ice it will be expelled to form a liquid of composition $NH_3 \cdot 2H_2O$ as soon as the temperature rises above 176 K, leaving a residue of pure ice. This process is called partial melting (because only a fraction of the solid actually melts) and in this case it happens nearly 100 degrees below the melting point of pure water-ice. A phase diagram to illustrate this effect is given in Fig. 2.6. The ice would stop melting once all the ammonia had been used up. The melt that had been formed would be free either to separate out and migrate independently, or to act as a lubricant between the ice grains, thus greatly reducing the effective viscosity of the ice. Even where volatiles are absent, the ice is likely to contain dissolved salts; so any melt will be a brine rather than pure water.

Many of the issues raised so far in this chapter will be discussed further in Chapters 4–7, taking advantage of the observational data that *Voyager* has provided. In the meantime, we conclude with a synopsis of the outer planet satellites as they were understood before the *Voyager* encounters.

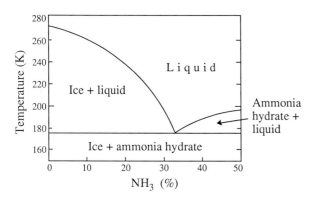

Fig. 2.6. Schematic phase diagram for a mixture of water and ammonia at low pressures. Water-ice and ammonia hydrate occur as a mixture of crystals at temperatures below 176 K. If the temperature rises above this, for a water-rich mixture (<30% ammonia), an ammonia-rich melt forms with the approximate composition $NH_3 \cdot 2H_2O$, leaving a residue of water-ice as the solid phase.

2.4 The pre-*Voyager* view

In the final years leading up to the *Voyager* missions, the observational evidence and the theoretical modelling discussed in this chapter led to a consensus rock-plus-ice model for the composition of the outer planet satellites, but there was debate about the degree of differentiation that they were likely to have experienced. Even if accretion had been homogeneous, it would be possible for a satellite to have a rocky core surrounded by ice if accretion, radioactivity, or tidal dissipation had generated sufficient heat. However, in the absence of any external heat sources, these bodies would have become solid through-out, leaving little scope for any kind of geological activity throughout most of their histories.

Debates of this sort could have gone on indefinitely, and in a fairly futile manner, without the *Voyager* project. Almost everybody was surprised by the variety in the surface histories of the outer planet satellites that was revealed on the *Voyager* images. This chapter has carried the analysis of the interiors of these bodies quite a long way, but it must be remembered that this discussion is based on physical and chemical models with limited observational support, mainly from crude spectroscopic data. Pictures of satellite surfaces played no role in the development of the basic understanding of their evolution as expressed so far, and we have done no more than touch upon the action of geological processes in shaping the faces of these worlds.

It is often overlooked that the *Voyager* spacecraft were not the first missions to the outer solar system. Back in the 1960s there was serious concern that there would be so much small debris in the asteroid belt beyond the orbit of Mars that any spacecraft attempting to pass through this region would be almost certain to collide with something and be destroyed. The pessimistic view went on to state that even if a spacecraft survived passage through the asteroid belt it would be disabled before it came close enough to Jupiter to be of any use because of the intense radiation concentrated by the planet's magnetic field. To see if it was feasible to explore the outer solar system, two spacecraft, *Pioneer-10* and *Pioneer-11*, were launched in 1972 and 1973, respectively. They were small by today's standards, only 258 kg apiece, of which the scientific instruments made up just 25 kg. Much to everyone's relief both of these spacecraft survived to give the first close-up pictures of Jupiter's cloud formations, and, more importantly for the future missions, reassuring data on the intensity of interplanetary particles and fields. After passing Jupiter, *Pioneer-11* went on to become the first mission to fly through the Saturn system.

The imaging capability of the *Pioneers* was crude and little priority could be devoted to obtaining pictures of the planetary satellites. Such pictures as were obtained were undoubtedly sharper than any view seen from Earth, but the sort of detail that they showed did little more than confirm the vague notion of streaky markings on Ganymede (Fig. 2.7).

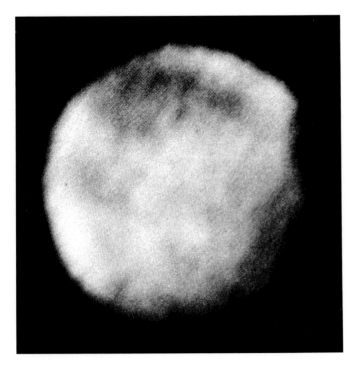

Fig. 2.7. The best image of Ganymede recorded by *Pioneer-10* in 1974. Gross variations in surface brightness can be seen, but no details are visible.

Nevertheless, these two pathfinder missions proved that it could be done, and the scene was set for an ambitious series of successor missions. This became the *Voyager* project, and because so much of our geological knowledge of the planetary satellites derives from this, it deserves a separate chapter.

3

Voyager

. . . far outside the flaming walls of the world . . .

Lucretius, *De Rerum Natura*

It is not easy to get to Jupiter and even harder to reach the planets beyond. Apart from the hazards posed by the risk of collision and radiation damage, the distances are very great and the journey times involved are so long that the spacecraft must be relied upon to continue to function and stay in radio communication with the Earth for several years. As Table 3.1 shows, Saturn is about twice as far away as Jupiter, and Uranus is twice as far again.

It took *Pioneer-10* almost two years to reach Jupiter. Bearing in mind that a spacecraft has to fight the Sun's gravity all the way, it would normally take considerably more than four times as long to reach Uranus. This is a long time to wait for results, especially as a mission is typically in the pre-launch planning stage for several years beforehand. To send a spacecraft on such a long journey to a single distant planet would also entail a large element of risk that it would break down *en route*, making the whole mission virtually a wash-out. By a stroke of immense good fortune, the *Voyager* spacecraft were able to minimize these problems by visiting several planets in turn while still reaching their final targets faster than if they had been sent by a more direct route.

Table 3.1 The scale of the solar system, showing the distances of Earth and the outer planets from the Sun. The values quoted are the length of the semi-major axis of the orbit. AU stands for astronomical unit, which is defined by the size of the Earth's orbit

Planet	Mean distance from Sun		Revolution period
	(10^6 km)	(AU)	(years)
Earth	150	1.00	1.00
Jupiter	778	5.20	11.86
Saturn	1426	9.54	29.46
Uranus	2868	19.18	84.07
Neptune	4494	30.06	164.82
Pluto	5900	39.44	248.6

3.1 The Grand Tour

It can be seen from Table 3.1 that each planet takes a different length of time to orbit the Sun. This means that their relative alignments are continually changing, as the inner planets overhaul those orbiting beyond them. By a lucky coincidence, during the 1970s the outer planets were positioned such that a spacecraft sent from Earth to Jupiter could carry on and visit Saturn, Uranus and then either Neptune or Pluto in turn without needing impracticably large amounts of fuel to change course. This meant that even if a mission to Neptune were put out of action half-way through its journey, it could already have achieved useful goals at the less-distant planets. In 1966, Gary Flandro, at that time a graduate student working at the NASA Jet Propulsion Laboratory, realized that a spacecraft could tour the outer solar system in this way by using the gravitational field of each planet it encountered to accelerate and speed it towards the next target on a gravity-assisted trajectory, more graphically known as the 'gravitational sling-shot' effect. Thus was born the concept of the Grand Tour, in which a single spacecraft would take in most of the sights of the outer solar system, needing about a decade to complete the mission. Such a planetary alignment recurs only once in every 180 years and it is particularly fortunate that space technology was sufficiently well developed by this time, otherwise there would have been a very long wait for the next such favourable opportunity. In the event, only the USA, through the agency of NASA, was able to mount an effort along these lines.

NASA's original plan was for a Grand Tour fleet of four spacecraft that would swing past Jupiter and thence, between them, to all the planets beyond, including Pluto. In the heady days early in the *Apollo* Moon-landing era the US Congress was willing to allocate NASA the funds needed to equip and launch these missions. However, NASA decided that its future priorities would be to the manned spaceflight programme, and in particular the development of the Space Shuttle, with the result that the Grand Tour concept was officially abandoned. A much trimmed-down version was kept on the books in the form of the *Mariner* Jupiter–Saturn project. This was funded from 1 July 1972 onwards, and the team was instructed to design and build two spacecraft that would have a fair chance of surviving encounters with both Jupiter and Saturn and to include an investigation of Titan. As NASA was not willing to commit its budget beyond the projected date of the Saturn encounters, there was to be no instrumentation nor electronics specifically designed for targets of opportunity beyond Saturn.

The project was subsequently renamed *Voyager*, and the spacecraft that were eventually built and launched became *Voyager-1* and *Voyager-2*. The nominal mission plan remained the same, and it was not until *Voyager-2* had passed Saturn (and executed a gravity-assist manoeuvre to speed it on towards Uranus) that funding for the extended project was approved. The trajectories of the two spacecraft are shown in Fig. 3.1.

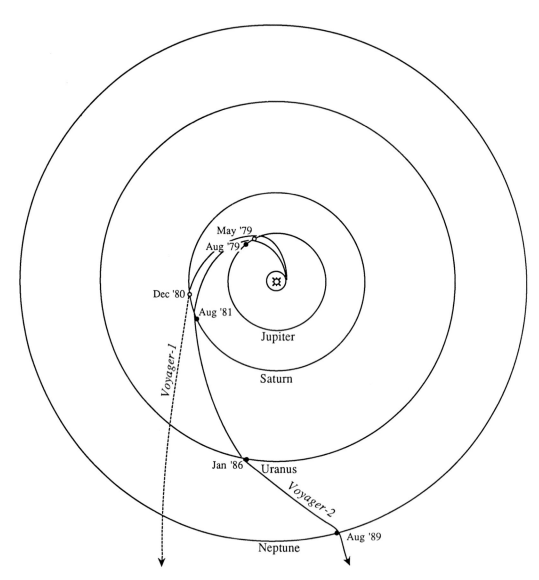

Fig. 3.1. Trajectories of the two *Voyager* spacecraft. *Voyager-1*'s encounter with Saturn flung it onwards above the plane of the solar system. After Neptune, *Voyager-2* continued on a course heading below the plane of the solar system.

3.2 The *Voyager* project

The *Voyager* spacecraft had several objectives. Foremost among these was gathering information about the masses, magnetic fields, composition, and atmospheric features of the outer planets. As conceived in the mission plans, the satellites were to be of secondary importance. In addition, the *Voyagers* were equipped to measure charged particles and magnetic fields in the long voids between planets. The two *Voyagers* were launched in 1977 (Fig. 3.2), *Voyager-2* on 20 August and *Voyager-1*, on a faster trajectory that allowed it to reach Jupiter before its twin, on 5 September.

Fig. 3.2. Launch of *Voyager-1* by a Titan-Centaur rocket from Cape Canaveral on 5 September 1977.

3.2.1 *The* Voyager *spacecraft*

The two *Voyager* spacecraft were identical, each weighing 825 kg and carrying 105 kg of instruments for eleven scientific investigations. The basic layout is shown in Fig. 3.3. Spacecraft designed to operate in the inner solar system usually draw their power from arrays of solar cells that convert sunlight into electricity. This was not a viable option for the *Voyagers*, because sunlight is too weak in the outer solar system; at Jupiter it is less than 4 per cent and at Neptune only 0.1 per cent of its brightness at Earth. Instead, the

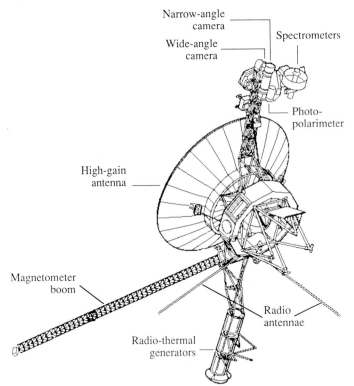

Fig. 3.3. Configuration of a *Voyager* spacecraft. Hydrazine thrusters for attitude control and trajectory correction were arranged around the ten-sided central mounting.

Voyagers were nuclear-powered, using the crude but effective technique of generating electricity from the heat produced in radioisotope thermal generators, which were mounted on a boom on the opposite side of the spacecraft to the sensitive scientific packages.

The cameras and other optical devices were mounted on a stabilized scan platform, which was designed to be fully pointable allowing a camera to look at its target but keeping the high-gain antenna aimed exactly at Earth. In addition to being able to point the cameras at a succession of targets while maintaining a radio link with Earth, the movement of the scan platform provided a means of keeping a camera pointing precisely at its target even when the rate of relative motion between the spacecraft and target would otherwise have smeared the image. Image-motion compensation of this sort is particularly important in the outer solar system, where the speed of the spacecraft is greater and the level of sunlight is less, necessitating exposures of longer duration (typically 15 s at Neptune).

Near the end of its Saturn encounter, *Voyager-2* suffered a mishap when one of the axes of the scan platform jammed during a pre-programmed slew, which led to an automatic power shutdown of the system. As soon as the problem was realized back on Earth, a fresh command was sent to the spacecraft to turn the scanner power back on, but the jammed axis refused to budge. This could have seriously prejudiced the imaging programme at

Uranus and Neptune, but in one of the *Voyager* programme's many triumphs of engineering compromise, the ground crew managed to free the axis by repeatedly warming it (using a heater intended to keep the lubricating oil at a suitable temperature) and then allowing it to cool again. This eventually freed the axis, but from then on high slew rates were banned by the science team and even medium-rate slews were limited to essential manoeuvres necessary to redirect the cameras to new targets. In the end, the image-motion compensation needed to obtain sharp pictures during the Uranus and Neptune encounters was achieved by swinging the spacecraft as a whole, using the attitude control jets. Fortunately there was plenty of fuel left for these, because the original navigation was so good that very little fuel had been needed for course corrections. An additional use of the attitude control jets, introduced at Uranus and refined for the Neptune encounter, was to fire them in a specific combination to compensate for the slight torque caused every time the tape recorder was turned on or off, thus removing an additional source of image smear and pointing inaccuracy.

Each *Voyager* spacecraft carried two cameras; a narrow-angle camera with a field of view just under half a degree across and a wide-angle camera with a field of view of 3.2°. Almost all the pictures in this book were taken with the narrow-angle camera. The wide-angle camera was used more for searching for unknown satellites, imaging planets when they were very close and obtaining navigational images of star-fields. Both cameras were vidicon devices, recording their image in a similar manner to a television camera. However, unlike television, the image was broken down into rows of discrete picture elements, or pixels, before being transmitted. Each image consisted of 800 rows of 800 pixels each, and the intensity of light recorded in each pixel could be anywhere in the range 0–255, a choice of 256 possible levels. Note that this number is 2 to the power of 8, or 2^8, which in computer terms is 8 bits of information. A number in this range is convenient to transmit as a string of code and to handle in an image-processing computer where the image is reconstructed and displayed. An enlargement of part of a *Voyager* image, showing the individual pixels, is reproduced in Fig. 3.4.

Digital images such as these can almost always be improved by image-processing techniques. Geometric distortions due to the camera can be compensated for by reference to calibration dots (reseau marks) that were etched on to the vidicon faceplate at precisely measured locations. Reseau marks and image defects such as bad scan-lines can be removed and replaced with synthetic pixels or synthetic lines by interpolation from the surrounding area. Contrast can be adjusted as appropriate for dark or bright regions of the surface, and fine details can be enhanced by spatial filtering techniques such as edge enhancement. The image can also be warped to fit a standard map projection if desired. Most of the images reproduced in this book have had the reseau marks removed, all have undergone contrast stretching, and some have been filtered to enhance certain details.

A vidicon is essentially a black-and-white imaging device, because it has no colour sensitivity. Colour, however, can be an important clue to the composition of surfaces and atmospheres, and so each camera included a filter wheel with a choice of coloured filters to accept light of specific wavelengths. The *Voyager* narrow-angle camera had six different filters: ultraviolet, violet, blue, green, orange, and clear (which allowed all wavelengths

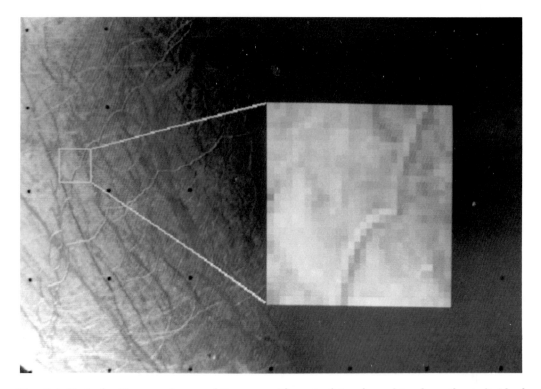

Fig. 3.4. Part of a *Voyager* image of Europa, with part of it enlarged to show the individual pixels (picture elements) from which it is constructed. The black dots in the main image are reseau marks (see text for explanation).

through). To obtain a colour picture, *Voyager* had to record separate images in each of three colours, which could then be combined back on Earth to give a coloured product. By combining orange, green, and blue the colours appear fairly natural. However, because of the time necessary to rotate the filter wheel to a new position, the viewing geometry had usually changed noticeably between frames, requiring two of the images to be warped to fit the third by geometric image-processing techniques.

As a *Voyager* probe travelled further away, the strength of its signals received on Earth became weaker, making it harder to transmit its images and other data without an unacceptable level of noise. While at Jupiter the transmission rate was 115 200 bits per second, but this fell to a maximum of 21 600 bits per second at Uranus and Neptune, despite increasing the size of the antennae used in NASA's Deep Space Network and electronically roping in some radio-telescopes to help reduce the noise. An image of 800 lines with 800 pixels each consists of 5 120 000 bits, so it would take about four minutes simply to transmit an image from Neptune. At that rate, bearing in mind all the other experiments going on during a planetary encounter, the on-board tape recorders would have filled up before all the desired images could have been handled. Another of the

Voyager project's technical triumphs was the reprogramming of *Voyager-2*'s on-board computer while it was travelling between Saturn and Uranus, which allowed it to compress the image data. Instead of transmitting an 8-bit number representing the brightness of each pixel, what was sent was simply the difference in brightness between successive pixels. This reduced the average number of bits to four per pixel, representing a reduction of the data volume by a factor of two without loss of information, except in particularly complex scenes where the allocation of bits per scan-line was sometimes used up, leaving blanks that had to be interpolated.

3.2.2 *The encounters*

In devising the *Voyager* encounters with the planets and their satellite systems, the mission planners had a near-impossible task. The main requirement was that the encounter should give the spacecraft the right gravitational boost to send it exactly on its way towards its next encounter. Among the other requirements were that the spacecraft should pass as close as possible to the planet and as many of its satellites as could be managed (to enable high-resolution imaging and other measurements), and yet avoid flying through any known rings. There was also a requirement to pass behind each planet as seen from Earth (so the attenuation of the radio signal from the spacecraft could be used to determine the structure of the planet's atmosphere), and also to pass behind any rings for similar radio occultation measurements.

During an encounter the spacecraft velocity was so high that there was no scope for slowing down or altering course by using the on-board thrusters, so ideally an encounter would be timed such that the satellites were arranged to let the spacecraft fly past each one in turn, rather like a miniature Grand Tour. In practice this could not be fully achieved, although between them the two *Voyagers* were able to cover the Jupiter and Saturn systems fairly well. As Fig. 3.5 shows, *Voyager-1* was able to pass close by Io, Ganymede, and Callisto, but not Europa, whereas *Voyager-2* passed by Callisto, Ganymede, and Europa, but had no opportunity to study Io closely. Similar trade-offs were made at Saturn, but less satisfactorily, as there were more satellites to take into consideration.

Passing close to a satellite is of limited use unless it is possible to obtain good quality images of most of its surface. How well the two *Voyagers* were able to cover the Jupiter system is indicated in Fig. 3.6, which shows the distribution of the most detailed images available. Fine resolution images cover only restricted areas, and Europa in particular still has large tracts that have never been seen in detail. The coverage of satellites in the Saturn system is patchy in the same way. The requirement that *Voyager-1* should fly behind Titan (for radio occultation measurements of its atmosphere) meant that it had to proceed onwards over Saturn's south polar region, whence the gravitational sling-shot effect swung it northwards and directed it away at an angle of some 35° above the plane of the solar system. This precluded any further planetary encounters, so only *Voyager-2* went on to visit the outermost planets.

There was not much choice about *Voyager-2*'s encounter with Uranus; at the time of the encounter (and as it will remain until around the end of this century) the axis of Uranus and

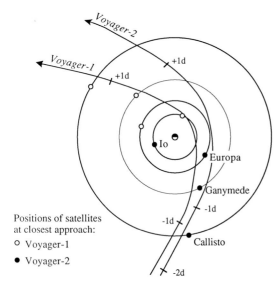

Fig. 3.5. *Voyager-1* and *Voyager-2* trajectories through the Jupiter system. Jupiter itself is shown to scale, but the sizes of its satellites have been exaggerated. The tick marks show the time (in days) before and after closest approach to Jupiter. *Voyager-1*'s closest satellite encounter was at a distance of 18 640 km from Io, and *Voyager-2* passed within 60 000 km of Ganymede.

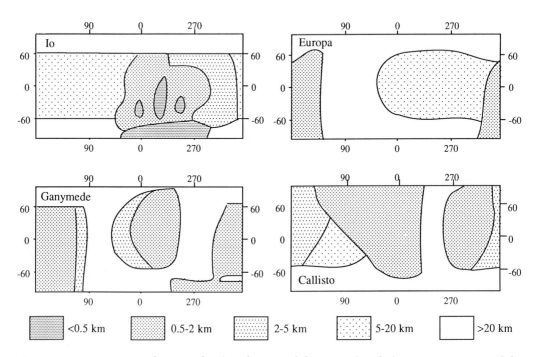

Fig. 3.6. Mercator maps showing the distribution of the most detailed *Voyager* images of the Galilean satellites. The resolution is expressed as the size of the smallest pixels in the images.

its satellite system was pointing more-or-less directly towards the Sun, the direction from which *Voyager-2* approached. This meant that *Voyager-2* had to be aimed close to Uranus, rather like a dart at a bull's-eye (Fig. 3.7a), with no opportunity to pass close to its outlying satellites on the way in or out. Fortunately, Miranda, the one satellite it was possible to arrange a close pass to because it was the innermost of the known satellites, turned out to be unexpectedly fascinating. Out at Neptune the situation was different again; there was no onwards targeting constraint, so *Voyager-2* was sent as close to both Neptune and Triton as possible, passing into occultation by these bodies at optimum times for signal reception on Earth (Fig. 3.7b).

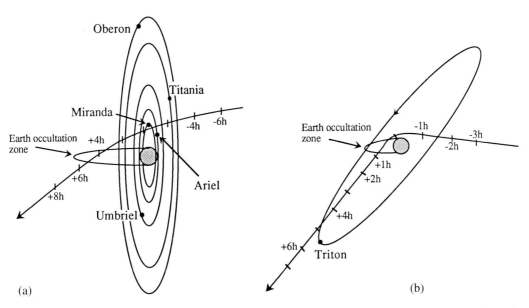

Fig. 3.7. *Voyager-2* trajectories through the (a) Uranus and (b) Neptune systems. The tick marks show the time (in hours) before and after closest approach to the planet.

3.2.3 *Onwards into the abyss*

Voyager-2's encounter with the Neptune system in August 1989 flung the spacecraft onwards at an angle of 45° below the plane of the solar system. After recording a final imaging mozaic sequence showing a view of the whole solar system from outside, the imaging system and other planetary experiments were turned off. However, the cosmic ray, charged-particle, and magnetometer experiments were left running, with the antenna transmitting on low power.

At the time of writing, both *Voyagers* and indeed both *Pioneer* spacecraft are still functioning. The *Voyager* radio-thermal isotope generators will probably give them enough power to continue functioning and to stay in contact with the Earth until about 2015–2020. The race is now on between these four spacecraft to see which is first to cross the heliopause, which marks the boundary between the Sun's magnetic field and the incoming stellar wind. The smart money appears to be on *Voyager-1*, and it is likely that this will have the honour of becoming mankind's first probe into the interstellar medium, some time towards the end of the 1990s.

3.3 Mapping the satellites using *Voyager* images

The *Voyager* images are the only data we have that can be used to map the surface features of the satellites of the outer planets. For most geological work these images can be used directly; for example, a geologist will be interested in investigating the overlapping relationships between different landforms to determine the geological history of the satellite, because younger deposits should overlie older units, and this information can be seen on the basic images. However, to fit all the information into a global framework, it is necessary to have a reliable map. Using the images to produce a map to cartographic standards is a long and complex procedure that is beyond the scope of this book. The responsibility of achieving this rests primarily with a team from the United States Geological Survey based in Flagstaff, Arizona. Among the tasks that have to be under-taken are interpreting the scale within each image from a knowledge of the relative positions of the satellite and the spacecraft and the pointing direction of the camera, deter-mining the best-fit spheroid to the satellite, and defining a co-ordinate system so that the data can be plotted on a standard map projection.

It is possible to make a mosaic of the corrected images and to warp them to fit a map projection using digital image-processing techniques, but a more common approach is to redraw the information in the images, using the skills of experienced airbrush artists. This enables the best details from several images of a single area to be combined, taking account of features that show up best under different illumination conditions or at different scales. In this way a large part of a satellite can be shown in a consistent fashion, and can be drawn so that the angle of solar illumination appears uniform throughout. A certain amount of artistic licence is involved in this process, but experience from Mars, which allowed airbrush maps based on early low resolution images (from the *Mariner* series) to be compared a few years later with higher resolution *Viking* Orbiter images, shows that the artists can usually be relied upon. An example of a shaded relief airbrush map is shown in Fig. 3.8.

The co-ordinate system for a synchronously rotating satellite is simple to define. The equator is, as always, 90° from the poles of rotation, and 0° longitude is defined to run through the centre of the planet-facing hemisphere. As a result of the inevitable uncertainties in determining the orientation of the images, the origin of the co-ordinate system is in practice defined relative to the positions of various topographic features that are visible on the images. To refer to these control points, and for general descriptive use, features on the surface have to be named.

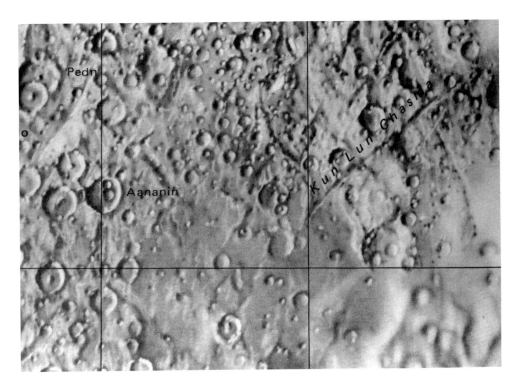

Fig. 3.8. Detail from a shaded relief airbrush map of Rhea. The lines are latitude and longitude at 30° intervals. The lack of detail near the south-east corner is because the resolution was poorer, or the view more oblique, on the best images of this area.

3.3.1 *The naming of names*

With several uncharted worlds to cover, the assignment of names is an immense task. To prevent the proliferation of haphazard, inappropriate, and duplicate names, planetary nomenclature is controlled by task groups of the International Astronomical Union. Each of the mappable outer planet satellites has been assigned a theme, usually based on mythology and sometimes connected with the myth from which the satellite's own name was drawn, although the satellite's discoverer is often honoured as well. Within the Uranus system there was already a strong theme set up in the names of the satellites themselves: Miranda, Ariel, Titania, and Oberon are all Shakespearean characters (in *The Tempest* and *A Midsummer Night's Dream*), and Umbriel (along with Ariel) appears in Pope's *The Rape of the Lock*. To continue this theme, the ten smaller satellites of Uranus that *Voyager-2* discovered were each given a name out of Shakespeare, and the names of surface features on the mappable satellites were also drawn from Shakespeare, except on Ariel and Umbriel where names of good or bright and evil or dark spirits, respectively, were chosen from mythologies from around the world. In devising names, the International

Astronomical Union has tried to avoid any cultural or racial bias throughout the solar system as a whole; for example, the names appearing on the map extract covering Rhea in Fig. 3.8 are drawn from creation myths of Asian, African, and South American origin. The naming conventions for the satellites discussed in this book are tabulated in the Appendix (pp. 187–8).

Craters, such as Aananir on Fig. 3.8, are given a straightforward name without qualification, but other features are usually given a Latin appendage that describes the general nature of the feature. An example of this sort is Kun Lun Chasma on Fig. 3.8, where *chasma* (plural: *chasmata*) means a canyon-like feature. Other common terms are: *fossa* (plural *fossae*), for a long narrow, shallow depression; *linea* (plural *lineae*), for an elongate marking; *mons* (plural *montes*), for a mountain; *patera* (plural *paterae*), for a shallow crater with a scalloped or complex edge; and *vallis* (plural *valles*), for a sinuous linear depression. The use of Latin terms may seem perverse, especially the confusing plural forms, but at least it avoids giving names which imply that we know the true nature and origin of the feature. For example, a possible interpretation of Kun Lun Chasma is that it is a graben, meaning a valley created when parallel faults allowed the floor to subside, but there are other possible explanations such as tidal fracturing, excavation by oblique impact, or volcanic fissuring. Thus it would be very rash (and unkind to future generations) to make its official name the Kun Lun Graben, and we avoid committing ourselves by the use of the purely descriptive term *chasma*. However, this practice does not mean that an effort should not be made to understand the origin and significance of the features observed on the planetary satellites, and it is to that end that most of the remainder of this book is devoted.

4

Icy lithospheres

... remember of this unstable world ...

Malory, *Le Morte D'Arthur*

With such a diversity of worlds among the satellites of the outer planets, you would be excused for supposing that they have too little in common for any mutual thread to run through a discussion of their histories. However, by adopting a geological stance it is possible to set up a unified approach to discussing these worlds, which at the same time helps us to understand the differences between them. To do this, we must look at depth-related changes in the physical properties of the material of which these worlds are made. In particular, changes from 'strong-and-rigid' to 'weak-and-plastic' states are crucial. By treating chemical composition as of subordinate importance, this approach will also illustrate the close similarities that the icy moons have with the Earth and the other terrestrial planets. Since we know most about the Earth, it is useful to start by summarizing the relevant aspects of the Earth's internal structure.

4.1 Inside the Earth

Just about everybody has heard that the Earth has a core overlain by a mantle, which in turn, is overlain by an outer skin known as the crust (Fig. 4.1). The existence of these principal units can be demonstrated by the way that seismic waves generated by earthquakes are reflected and refracted at the interfaces between them. There are subdivisions within the mantle marked by changes in the speed at which seismic waves travel, and the core is recognized as having a fluid outer part because this transmits pressure waves but not shear waves.

The core is distinct from the mantle both physically and chemically. It contains metallic iron and nickel, and probably sulphur, which is chemically unable to enter into silicate minerals. The dense nickel–iron mixture is presumed to have segregated gravitationally towards the centre after being liberated by melting due to accretional and radiogenic heating early in the Earth's history.

Two of the most common elements in the Earth are silicon and oxygen. An atom of silicon bonds readily with four oxygen atoms to form a tetrahedral-shaped silica unit. This is the fundamental building block of all the common minerals in the crust and mantle. A silicon atom can share one or more of its oxygens with another silicon atom, in which case the silica tetrahedra form chains, sheets, or more complex three-dimensional structures. The balance of chemical charges is kept neutral by the incorporation of metallic ions into

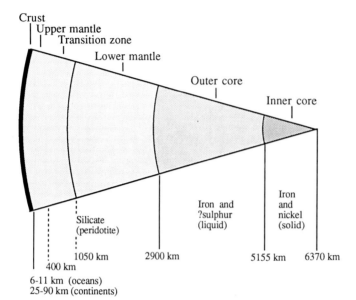

Fig. 4.1. The compositional layers within the Earth.

the structure; in the mantle the most abundant of these are iron and magnesium. The mantle, then, consists essentially of iron and magnesium silicates. It is thought that its chemical composition (essentially that of the rock known as peridotite) does not change significantly with depth and that the divisions within the mantle shown in Fig. 4.1 correspond to phase changes, where the crystalline structure rearranges to denser forms in a manner analogous to the ice phase changes discussed in Chapter 2. Even on the Earth, geologists have been able to obtain samples only from the upper part of the mantle, and the exact nature of the deeper phase changes is a matter of speculation.

The crust is the bit we live on and its significance in terms of the planet as a whole tends to be exaggerated. It is, after all, very thin compared with the total size of the Earth. The crust was formed after accretion by partial melts from within the mantle that rose to the surface and solidified. As a result, it is chemically distinct from the mantle, being richer in silica and containing important amounts of elements other than iron and magnesium, which combine with silica to form a new set of minerals. These give the crust a lower density than the mantle, and it is this density contrast that is responsible for the clear seismic division between them, known as the Moho, or Mohorovičič discontinuity.

So far as can be told, this description of the Earth's layered structure is moderately representative of the other terrestrial planets, especially Venus, whose size and density are much the same as the Earth. There is a complication in the case of the Earth, in that it has two quite distinct types of crust. Oceanic crust is basaltic in composition, that is, with several per cent more silica than in the mantle and with elements such as aluminium and calcium being comparable in abundance to magnesium and iron. Continental crust is more granitic in composition, having even more silica than oceanic crust, and with significant concentrations of sodium and potassium, but correspondingly less iron and magnesium.

Generally, oceanic crust is about 6–11 km thick, whereas continental crust varies between 25 and 90 km in thickness.

4.1.1 *The Earth's lithosphere and asthenosphere*

For studying all except the geochemical aspects of the Earth's geological history (and similarly for the other planets), the distinction between crust and mantle is of only secondary importance. To understand the operation of the processes responsible for the deformation of the surface and the volcanism that has resurfaced much of it, we must look instead at how the planet changes in strength with depth. Mechanically speaking, the uppermost part of the Earth's mantle is entirely coupled with the crust. If the crust is displaced, then so is the immediately underlying mantle. Together, the crust and upper-most mantle comprise what has become known as the lithosphere, using the Greek root *lithos* (meaning rock). This outermost layer of the solid Earth is rocky in the sense that it is rigid and brittle. Its total thickness is 100–200 km below the continents and 100 km or less in the oceans (Fig. 4.2). What enables parts of the lithosphere to move around and interact with each other is the fact that directly beneath it is a much weaker layer that flows plastically in response to stress and that will convect if there is an adequate thermal gradient. This layer is termed the asthenosphere, using the Greek root *astheno* (meaning weak).

The Earth's asthenosphere is weak because the local pressure–temperature conditions enable the material of which it is made to deform in a near-fluid fashion over geologic time. The asthenosphere does not have a well-defined base, and the mantle may well be convecting in the solid state all the way down to the core–mantle boundary. The situation is compounded by the fact that in many areas of the Earth the top of the asthenosphere contains a small fraction of melt (i.e. it is partially molten), which contributes to the general

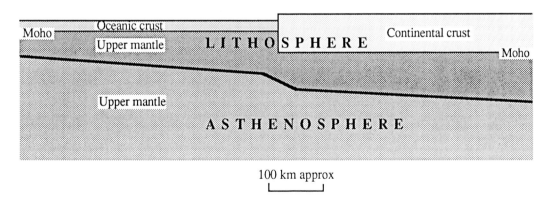

100 km approx

Fig. 4.2. The most important layers of the Earth in terms of its mechanical properties; the rigid lithosphere and the plastic asthenosphere, showing also the difference between continental and oceanic crust. The mantle below the uppermost part of the upper mantle belongs to the astheno-sphere. In most areas the top of the asthenosphere includes a partially molten layer known as the low-velocity zone.

weakness of the asthenosphere by lubricating the motion between solid grains. This partial melting is another consequence of the pressure–temperature conditions, which encourage certain elements to be squeezed out of the crystals in which they were accommodated and into a liquid phase. If any of this melt reaches the surface, it gives rise to volcanoes. The partially molten zone is recognized by the anomalously low speeds at which it transmits seismic waves, and consequently it is often referred to as the low-velocity zone.

The existence of the Earth's asthenosphere was first postulated to account for the isostatic compensation of mountain ranges, such as the Himalayas, where the crust is anomalously thick. The crust rides buoyantly on the mantle, so mountain ranges are (almost literally) only the tip of the iceberg, with most of the thickening of the crust occurring below ground level. To allow the crust to project downwards into the mantle, there must be a weak layer, the asthenosphere, somewhere near the top of it. This principle of isostasy also explains why oceanic crust, being thinner and denser than continental crust, is lower-lying. Another demonstration of the existence of a weak layer, which incidentally enables estimates to be made of its strength and rate of flow, is provided by the fact that much of Scandinavia is still rebounding upwards at a rate of several millimetres per year in response to the removal of the major ice sheet that covered it during the last glaciation, but which melted away about 15 000 years ago. As we shall see later, similar reasoning can be used to infer the strengths and thicknesses of the lithosphere and asthenosphere on icy moons where major topographic features appear to have undergone isostatic adjustment since their formation.

Equally important on the Earth is the role of the asthenosphere in allowing the sideways motion of the lithosphere required by plate tectonics. Since the 1960s it has become almost universally recognized that the lithosphere is broken into plates that spread apart at oceanic spreading axes, such as the Mid-Atlantic Ridge and the East Pacific Rise, and are destroyed at subduction zones, where the plate motion converges and one plate dives down below another, such as beneath the Andes. It used to be thought that these plate motions were driven directly by convection within the mantle, perhaps within an asthenosphere confined to the upper mantle alone. More recently it has become clear that convection extends down deep within the mantle as well, but that this is not simply related to the pattern of plate motions.

4.1.2 Mantle convection

This deep internal convection occurs in the solid state, in the same manner that high-pressure phases of ice will flow, given sufficient time. To put this in context, the rates of flow in the top of the asthenosphere must be much the same as the speed of motion of the overlying plates, which approaches 10 cm per year. The speed of convection currents deeper in the mantle may be less, perhaps around 0.1–1 cm per year.

The important difference between the lithosphere and the asthenosphere is that convection can occur in the latter, but never in the former. Modern estimates of the viscosity of the lower mantle put it at about two to three times that of the asthenospheric upper mantle, so for many purposes the asthenosphere can be regarded as including the whole of the mantle below the lithosphere. The lithosphere responds to stresses applied from below primarily by fracturing. These stresses may cause it to be deformed in a ductile fashion at depth, but the lithosphere *never* acts as a fluid, whereas on a geological time-scale this is exactly how the asthenosphere behaves, and it has a crucial role in allowing the outwards transfer of heat by convection.

4.1.3 *Lithospheres of the other terrestrial planets*

The lithospheric thickness of the other terrestrial planets depends on how pressure and temperature increase with depth. The lithosphere on Venus is probably thinner than on Earth, as a result of the much higher surface temperature. Indeed, its surface has regions that appear to be young fold and collision belts, consistent with convergence between comparatively thin lithospheric plates, and tracts cut by closely spaced faults or fractures suggestive of fracturing in a thin lithosphere.

The remaining terrestrial planets have much thicker lithospheres than the Earth, because of their smaller size and consequently greater heat loss since they were formed. For example, Mars has bigger volcanoes than the Earth and Venus, but they are much fewer in number, suggesting that on Mars it has been harder to open a pathway for melts to penetrate upwards from the partially molten interior. In all provinces but one, volcanic activity ceased about 2.5 billion years ago, and the youngest province appears to have last erupted around 200 million years ago. Taken together with the lack of evidence for plate motions, this points to a lithosphere that was always thick and has grown thicker with time until it is virtually impossible for convection within the sub-lithospheric mantle to fracture the lithosphere and allow the access of melts to the surface.

The Moon is the only body other than the Earth for which there are adequate seismic data to detect the internal layers directly. It has a distinct crust forming the top part of a lithosphere that extends to a depth of about 1000 km, about two-thirds of the way to the centre. It seems clear that the Moon's lithosphere has been thickening with time, as a result of cooling, and that it was considerably thinner around 3.4–3.0 billion years ago when most of the major impact basins were flooded by the volcanic eruptions that produced the lunar maria. Mercury probably also has a substantially thick lithosphere today. It bears 3-billion-year-old scars of an episode of global compression, probably associated with internal phase changes due to cooling, but shows no signs of deformation or volcanism since then.

Thus lithospheres and asthenospheres can be recognized in the geological history of all the terrestrial planets. The question is, how can this Earth-derived terminology be applied to the moons of the outer planets, and how can it be used to help us understand their histories? Clearly the chemistry is very different, but, as we shall see, the physical principles are much the same.

4.2 The lithosphere of an icy world

We will leave Io aside until Chapter 7; it is effectively a terrestrial planet, both in terms of density and size, so its lithosphere and asthenosphere can be defined as for the Earth, although the main source of heat is tidal. For now we will attempt a generic treatment of the other moons. Variations on this model, as manifested by the differences in their histories, will be emphasized in the subsequent chapters.

4.2.1 Compositional layers in icy moons

The first stage is to see if the three-fold compositionally based division into crust, mantle, and core can be defined. The masses of icy moons suggest that they probably have far too little nickel and iron for this to be a potential core-forming phase, as in the Earth. However, referring back to Fig. 2.3, which shows the three stages in the evolution of an icy moon, you should be able to recognize core formation by inwards segregation of a dense component in much the same way as was described for the Earth. The difference is that in an icy moon it is the rock (i.e. the silicates) that has segregated to the centre, but it is a good analogy with the Earth to refer to the rocky central part of such a body as its core. The middle layer (pure water or pure ice) would then constitute the mantle, and the outer layer (an undifferentiated ice–rock mixture) would be the crust. The crust in the partiuclar example illustrated is relatively much thicker than the Earth's crust, but there is a more fundamental difference in that its origin is due to a completely different process. The Earth's crust is a result of chemical segregation from the mantle, whereas in this icy moon model the crust has the composition of the original, undifferentiated, mixture from which the moon formed. Furthermore, the mixture of rock and ice forming this crust will be denser than the icy mantle. The reverse is true on Earth, where the crust is less dense than the mantle (which is why it is rare to find mantle rocks exposed at the surface). However, on an icy moon such as this, if the crust is thin enough and the mantle is weak enough it should be possible for the crust to founder, squeezing mantle material, or melts derived from it, up to the surface. Thus, Fig. 2.3 is a model of an icy moon where it is possible to distinguish core, mantle, and crust, although there are important differences in the origin and relative densities of the crust and mantle compared with the terrestrial planets.

This model of the structure of an icy moon is a fairly middle-of-the-road option; the extreme alternatives were shown in Fig. 2.5. On the undifferentiated alternative there is clearly no division into core, mantle, and crust, although, we we shall see shortly, this does not preclude the existence of a lithosphere and asthenosphere. On the fully differentiated model, there is clearly a rocky core, just as in the Fig. 2.3 example, but because melting in this instance proceeded all the way to the surface, there is no surviving undifferentiated layer, and thus no crust if the definition of the previous paragraph is used. A world like this can be considered to have a core and a mantle only.

The fully differentiated alternative in Fig. 2.5 could equally well be the result of homogeneous or heterogeneous accretion, whereas the undifferentiated structure could result only from a homogeneous accretion process. To complicate the range of possible structures still further, there is an important variant of differentiation in a homogeneously accreted moon that must be noted. This is illustrated in Fig. 4.3. In this model the build-up of accretional heat is initially so slow and inwards conduction so poor that the centre of the body does not melt, and remains a mixture of rock and ice. As it grows in size, the amount of accretional heat produced and retained becomes sufficient to melt the ice in the entire outer part of the moon, all the way to the surface. As a result, this outer part differentiates, with the rock segregating downwards and the water forming the outer layer. The downwardly segregating rock fragments are brought to a halt when they meet the still-solid rock and ice mixture in the centre of the body. The immediate result is therefore a three-layered structure with rock and ice in the centre, overlain by rock only, overlain in turn by water only (which freezes to become ice). The rock layer is denser than the rock and ice that it overlies, and so it is gravitationally unstable. If the central rock and ice can be made mobile in any way, perhaps through a small amount of melting, but equally well by solid-state convection, there will be a strong tendency for the overlying rock to swap places with the rock and ice beneath, so the final structure will be a core of rock, overlain by a mantle of rock and ice, with the thick icy shell remaining on the outside. This model is complicated, but is attractive in many ways, not least because the style of segregation of the core from the mantle resembles one of the more likely mechanisms by which the Earth's core is thought to have formed.

The range of possible compositional layers of an icy moon as summarized above is shown in Fig. 4.4. The moon can be a body with a core that is either rock and ice or just rock, a mantle that is either just ice or rock and ice, a crust that is rock and ice or just ice, or maybe even no compositional differentiation at all throughout the entire body.

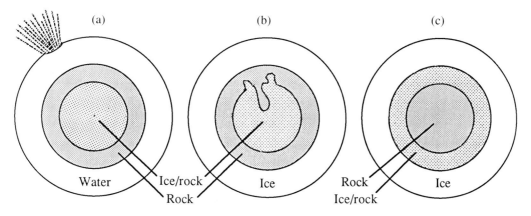

Fig. 4.3. Differentiation of a homogeneously accreted icy moon in which melting does not occur at the centre. (a) Towards the end of accretion the outer part has melted, allowing the rocky fraction to segregate inwards. The central part remains an undifferentiated ice–rock mixture. (b) Gravitational instability causes the denser rocky layer to displace the ice–rock mixture from the centre, to form in (c) a core of rock surrounded by an ice–rock layer, overlain by an outermost icy layer. The time difference between stages (a) and (c) is probably less than half a billion years.

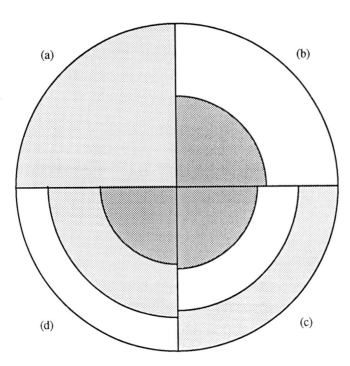

(a)

(b)

(d)

(c)

Fig. 4.4. The range of possible compositional layered structure in an icy moon. (a) Undifferentiated ice–rock mixture. (b) Fully differentiated with a rocky core overlain by ice. (c) Partially differentiated with a rocky core, an icy mantle and an ice–rock crust that has never melted. (d) Differentiated, with an ice–rock mixture overlying a rocky core and ice forming the outer layer; this could result from foundering of the crust in (c) or gravitational overturn as shown in Fig. 4.3. No scale is implied; in fact (a) is more likely in a smaller moon, (b) in a larger moon, (c) in a medium-sized moon, and (d) in either a large or a medium-sized moon.

In addition to internal differentiation, the outermost region of an icy moon that has remained otherwise undisturbed for hundreds of millions of years is likely to have both its physical and chemical nature altered as a consequence of the continuing impacts of meteorites and micrometeorites, and exposure to radiation. Some of the effects of meteorite impacts will be discussed in the next chapter; essentially impacts break the surface up into a fragmentary debris known as regolith, and because ice is volatilized more readily than rock, the surface may very gradually become enriched in silicates. However, these are superficial processes, of no great importance in determining global structure.

4.2.2 Ice-dominated lithospheres and asthenospheres

Ask them, and most people will tell you that ice is rigid. Hit it with a hammer and it will shatter. Ask a glaciologist, however, and you will get a different answer. Ice, you will be told, deforms under its own weight. In the form of a glacier, ice can flow downhill at a rate of several metres per year. Clearly, the distinction between rigid and plastic behaviour depends on the time-scale under consideration. This means that in an icy moon, whether the ice at a certain depth is regarded as belonging to the lithosphere or to the asthenosphere depends on the rate of strain that the ice is forced to respond to. In geological terms, the relevant strain rate is that which would be associated with flow at a rate of the order of 1–10 cm per year, at which speed the rate of heat transport by convection exceeds that of heat transport by conduction.

So how thick does this make the lithosphere of an icy moon? The ability of ice to deform in the solid state depends very strongly on temperature, so the answer to this question depends on the surface temperature and the thermal gradient (i.e. how quickly the temperature increases with depth). As a rule of thumb, on a geological time-scale ice acts in a fluid fashion, flowing by solid-state creep, if its temperature exceeds about six-tenths of the temperature at which it would melt. The equilibrium surface temperature of the moons of Jupiter is about 100 K, dropping to about 80 K out at Saturn. This means that near the surface of an icy moon water-ice is far below its melting temperature (273 K in the absence of confining pressure), and is well within the field of rigid behaviour, even at immensely long time-scales. Ice under these conditions is truly rock-like, which is the main reason why it is geologists in general and not glaciologists who can profit from the study of icy moons. Their low surface temperatures also explain why steep topographic features, such as crater walls, can retain their pristine form, without slumping downhill as would happen on Earth where ice is warmer and therefore able to deform under its own weight much more easily. Fig. 4.5 shows in graphical form how the change from fluid to brittle behaviour in ice varies with temperature and strain rate.

The temperature inside any planetary body increases with depth, and so the base of the lithosphere in an icy moon is reached at whatever depth the temperature reaches about sixth-tenths of the melting temperature of the ice. As an idea of scale, on Europa (a dense, massive satellite with a history of significant tidal heating, continuing to the present day) the lithosphere is probably no more than 30 km thick, whereas on Callisto (bigger and more massive than Europa, although less dense and without tidal heating) the lithosphere today is thought to be about 500–1000 km thick. Many of the moons of Saturn, which are considerably smaller, may at present be lithospheric throughout, unless the ice is made more fluid by the presence of contaminants (see later). As the icy moons have been cooling since their formation, the thicknesses of their lithospheres must have been growing at the same time. We shall see plenty of evidence of this in subsequent chapters.

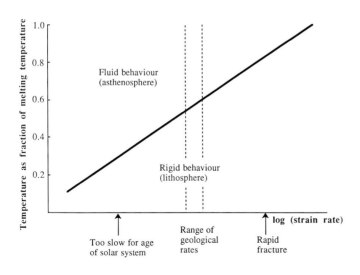

Fig. 4.5. Schematic representation of how the temperature at which solid-state creep can occur in ice depends on the rate at which the ice is deformed. Note that the strain rate is on a logarithmic scale. Temperature is expressed as a fraction of the melting temperature, which varies according to pressure (see Fig. 2.4).

The discussion so far as been concerned with the properties of pure water-ice, but it is not entirely reasonable to assume that this is the composition of either the lithosphere or the asthenosphere of an icy moon, because there are likely to be various contaminants mixed in with the ice. The most common, which is likely to be found in icy moons throughout the solar system, is rock. Fragments ranging from dust-size to house-size and above may be evenly distributed or concentrated in a particular layer depending on the differentiation history of the moon (Fig. 4.4). Rock fragments dispersed within ice increase its rigidity at a given temperature, and so will tend to make the thickness of the lithosphere greater than in a pure ice model. Counterbalancing this is the fact that because rock is denser than ice, the pressure in a rock and ice lithosphere must increase more rapidly with depth than within a pure ice lithosphere. As Fig. 2.4 shows, the melting temperature of ice decreases with pressure up to about 1 kbar (which is equivalent to a depth of about 180 km in Callisto and is not reached at all in moons smaller than Iapetus), so this pressure effect makes the ice itself less rigid at a given temperature. In general terms, then, dispersed rock fragments can make an icy lithosphere either thicker or thinner depending on a complex balance of effects.

Another important contaminant is ammonia (NH_3), which is likely to have been able to condense from the solar nebula in significant amounts at Uranus and beyond, and possibly even at Saturn. Ammonia and water can mix in any proportion within ice, but cosmic abundances suggest that the ammonia could at most make up no more than about 20 per cent of the total mass of the ice in planetary satellites. The principal effect of mixing ammonia-ice with water-ice is to make the ice more rigid, by hindering deformation within grains. This is essentially a lithospheric process, and it does not act to increase the viscosity of the asthenosphere because, as described in Chapter 2, ice that is contaminated by ammonia undergoes partial melting to produce an ammonia hydrate liquid of composition $NH_3 \cdot 2H_2O$ if the temperature rises above 176 K. This leaves a residue of pure water-ice with an intergranular fluid of the ammonia hydrate melt. If this melt remains at depth, as an intergranular fluid, then it would be analogous to the partially molten low velocity zone within the Earth's asthenosphere.

At Uranus and Neptune there may also be other contaminants in the ice, notably methane (CH_4), carbon monoxide (CO), or nitrogen (N_2). For every six molecules of water in ice I (but not the higher pressure forms), there is a void space within the crystalline lattice that is large enough to hold a molecule of any of these, in a structure known as a clathrate. This will have a poorly understood effect on the rigidity of the ice, but it seems clear that if the temperature rises sufficiently to drive the contaminant out of its lattice site and into a liquid phase, even if the contaminant occurs only in small amounts, this intergranular fluid must greatly reduce the effective viscosity of the ice as a whole, thus driving the lithosphere–asthenosphere boundary closer to the surface.

4.2.3 *Icy volcanism*

We have seen that icy moons can be layered in both compositional and mechanical senses. The ice in the asthenosphere is capable of flow at rates sufficient to sustain convection.

What about the production of fluids that could behave in a similar way to molten rock on the Earth? In other words, are 'igneous' processes possible?

We shall come back to specific examples in Chapters 6 and 7, but for now it is sufficient to remark that there are indeed many analogues to terrestrial volcanic processes. Owing to the low temperatures involved, eruptions on an icy moon are sometimes described as 'cryovolcanic', but this should not be allowed to confuse the essential similarities. A melt of $NH_3 \cdot 2H_2O$ formed by partial melting of water-ice contaminated with ammmonia will probably be slightly less dense than the remaining solid ice, and will definitely be less dense than the solid if this contains rock fragments. Such a melt would therefore tend to percolate upwards, but of course it would solidify as soon as it reached a region where the temperature dropped below 176 K. However, if a sufficient amount of such a melt accumulated in a pocket below the lithosphere before migrating up through a crack, *en masse* and rapidly, it could reach the surface before it froze. This melt would have physical properties (notably viscosity and yield strength) that under the low surface gravity of an icy moon would cause it to behave in a manner very similar to a basaltic lava flow on the Earth. In contrast, if the melt had undergone a significant amount of crystallization during its ascent, its viscosity and yield strength would be increased into the range exhibited by more silica-rich terrestrial flows, such as dacites and rhyolites. Thus ammonia–water melts provide scope for a variety of icy volcanic landforms that have terrestrial volcanic analogues.

Volcanism in the form of melts of pure or salty water is less easy to achieve for several reasons. One is the much higher temperature at which water or brine melts and at which it would freeze again *en route* to the surface. A related problem is that solid-state convection in water-ice can transport radiogenic or other heat outwards at a sufficient rate that the temperature is prevented from reaching the melting point. Another is that water is denser than ice and so it would not tend to rise, unless the ice were weighed down by rocky impurities. Any water-dominated magmas reaching the surface of an icy moon are likely to be very rich in crystals, that is, slush, or could even be ice mobilized by trace amounts of intergranular volatile fluids.

Another way of transporting the products of melting and mobilization at depth to the surface is by gas-driven processes, comparable with pyroclastic volcanism on the Earth.

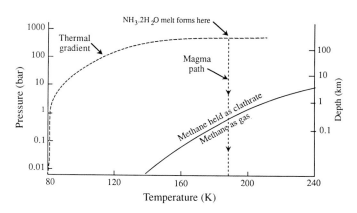

Fig. 4.6. A possible thermal gradient within an icy moon, showing the depth at which an $NH_3 \cdot 2H_2O$ melt could form and the path this magma would follow if it were to ascend rapidly without cooling. The solid line shows the conditions under which a methane clathrate decomposes to water-ice and methane gas. Thus the passage of the magma could liberate methane gas from the walls of the crack, resulting in an explosive eruption. The relationship of depth to pressure depends on the size and mass of the satellite concerned; the depth scale here is for Iapetus.

Trace amounts of ammonia in water could vaporize as the pressure decreased near the surface, or the passage of an $NH_3 \cdot 2H_2O$ melt through ice containing methane or some other volatile in clathrate form could cause the clathrate forming the walls of the crack to decompose into water-ice and gas (Fig. 4.6). Both these processes would drive the melt to the vent in an explosive fashion, whence it could be distributed either as frozen shards in a gas-driven jet or as a frothy liquid.

Now that we have examined the general physical and chemical factors that control the geology of the icy moons, we are in a position to examine them individually in the following chapters, to see what tales they have to tell about the forces that have shaped their surfaces.

5

Dead worlds

Strike flat the thick rotundity of the world
Shakespeare, *King Lear*

Rather than describe the moons of the outer planets one planet at a time, they are presented here in classes defined by the extent to which each exhibits evidence of geological processes. This chapter describes the worlds that show essentially no signs of geological activity on their surfaces except those associated with and recorded by impact cratering. As introduced in Chapter 1, heavily cratered surfaces must be older than less cratered areas because the latter can be assumed to have once been heavily cratered themselves and to have had their craters obliterated by some kind of resurfacing process. Chapter 6 discusses those worlds that exhibit clear signs of geological processes in the form of tectonism or volcanism that has overprinted and partly destroyed the earlier cratered terrains, but where such activity evidently belongs to a past era. Chapter 7 covers those worlds where this sort of geological activity is continuing today. Finally, Chapter 8 hazards some predictions about what conditions might be like on the few icy worlds of whose surfaces we have no good images.

The satellites included in this chapter are: Callisto (Jupiter); Mimas, Rhea, and Iapetus (Saturn); and Umbriel and Oberon (Uranus). We will take Callisto as our starting point as it is the most fully (and longest) studied world in this class. Almost everthing we know of the geology of these worlds is derived in one way or another from the study of their craters, but before delving into cratering processes in the outer solar system it would be best to summarize what is known about the production of craters closer to home.

5.1 Impact cratering

The planetological importance of understanding cratering first became apparent after studies of the Moon by Eugene Shoemaker and his colleagues in the early 1960s. These studies demonstrated a relationship between the relative age of stratigraphic units on the Moon as deduced from superposition (younger units overlying older ones) and the number of craters per unit area of surface (i.e. the 'crater density'). Older units have a denser distribution of craters. Since then, the stratigraphic units have been fitted into an absolute time-scale by the radiometric dating of samples returned to Earth by the six *Apollo* manned landings (USA) and the three *Luna* unmanned sample-return missions (USSR) that took place between 1969 and 1976.

Relative dating of the terrain units on the airless world of Mercury and the thinly atmosphered planet Mars can also be performed using cratering statistics. Their crater size distributions resemble those on the Moon, and dynamics arguments suggest that a population of impacting bodies whose orbits brought them into the inner solar system would become fairly evenly scattered as a result of gravitational interactions. In fact, although there is as yet no direct calibration from actual samples, the cratering time-scale set up on the Moon can be applied with reasonable confidence throughout the inner solar system, even on Venus, where craters tend to be destroyed by renewal of the surface by volcanic, tectonic, erosional, and sedimentary processes over comparatively short periods of time. However, a variety of evidence suggests that the bodies responsible for the cratering of the moons of the outer planets were not part of the same family, or 'population', that bombarded the inner solar system. To elaborate on this, it is necessary to delve into the abstruse realm of crater statistics.

5.1.1 Crater statistics

The Moon is dominated by two quite distinct terrains: the ancient highlands and the younger basaltic maria that resulted from volcanic flooding of most of the major impact basins. The oldest highlands are extremely densely cratered because they were subjected to bombardment by debris left over from the formation of the solar system, but most of this had been mopped up by the time the maria were formed, and these areas consequently have a considerably lower crater density. The contrast between the two terrain types is apparent in Fig. 5.1.

Simple visual comparisons are not sufficient to extract all the information that is lurking within images such as this. If we want to use craters to date a surface precisely or to tell whether a particular world was bombarded by the same population of impactors as others, it is important to quantify both the numbers and sizes of the craters present. The standard way to do this is to count the number of craters per unit area of surface whose diameters fall into successive size increments. The most straightforward way to display the results of this counting is to plot an 'incremental size–frequency distribution curve'. On such a graph, the crater diameter is plotted on the horizontal axis and the vertical axis shows the number of craters per square kilometer with a diameter that falls within each incremental range.

An example of size–frequency distribution curves for lunar highlands and lunar maria is shown in Fig. 5.2. Both the axes are on a logarithmic scale in order to cover a wide range of crater sizes and the vastly greater frequency of smaller craters than larger ones. This graph shows that the lunar highlands have a crater size–frequency distribution similar to that of the maria, except that the highlands have about a hundred times as many craters of a given size per unit area. If the rate of impacts had remained uniform through time, this would mean that the highlands must be about a hundred times older than the maria surfaces, but we know from radiometric dating that the highlands are around 4.0 billion years old and that the maria range from about 3.9 to about 3.1 billion years in age. This demonstrates a dramatic decrease in the cratering rate at about 3.9 billion years ago, marking the end of the late heavy bombardment period. Crater densities and radiometric ages on young

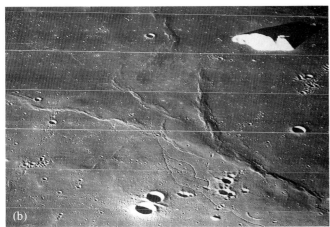

Fig. 5.1. Two parts of the Moon's surface photographed by an unmanned *Lunar Orbiter*, shown at similar scales. (a) Highland terrain, virtually saturated with impact craters inherited from the late heavy bombardment of the inner solar system. (b) A darker mare surface, which, being younger, has far fewer craters. An isolated peak of highland material rises above the lava-flooded mare at the top right.

ejecta blankets on the Moon suggest that the rate of impacting has remained fairly constant since at least 3.5 billion years ago.

If a surface had been bombarded by a single population of impactors consisting of few large objects but many small ones (as is implied by the known present distribution of potential impacting bodies in the Earth's vicinity), then we should expect a roughly straight line on an incremental size–frequency plot, as indeed is seen in Fig. 5.2. The gradient is typically about -2 in the inner solar system, except that it steepens to -3 or -4 for craters smaller than a few kilometres in diameter, many of which are secondary, having been produced by major blocks of ejecta from primary impacts. In practice, the gradient can become artificially less at the small-crater end of the distribution owing to a variety of circumstances, including the small craters being harder to see on the images, their being more easily buried by ejecta from subsequent larger impacts, and the surface becoming saturated with small craters so that for each one created there is, on average, a similar one destroyed. There are also subtle inflexions of the curve when a second population of

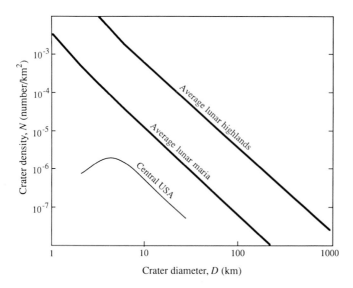

Fig. 5.2. Incremental size–frequency distribution curves for the two major terrain divisions on the Moon. There are many more small than large craters, and the lunar highlands have approaching a hundred times more craters than the maria. Variations in crater density within these major units (not illustrated) are used for relative dating. A curve for the central USA is also shown; there are many fewer craters because the surface is much younger, and the curve turns down at small crater sizes because small craters have been removed by erosion.

impactors with a different size–frequency distribution has partly overprinted the record of the earlier bombardment.

These effects are hard to see on an incremental size–frequency plot. The standard technique to overcome this is to plot how far the observed distribution departs from a line with a gradient of −3, on what is known as a 'relative size–frequency distribution plot' (Fig. 5.3). This has crater diameter on the horizontal axis as before, but shows a parameter R on the vertical axis, which is the ratio of the observed distribution to the function

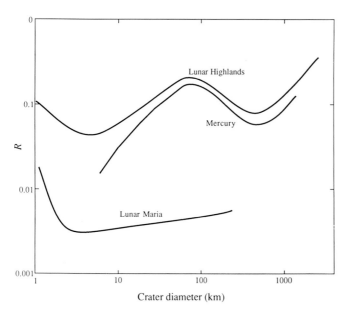

Fig. 5.3. Relative size–frequency distribution plot showing the differences in the distribution function for craters on the lunar highlands and the lunar maria. This indicates that the population of impactors that produced most of the highland craters was different to the later population that produced all the impact craters on the maria (which are younger than the highlands). However, there is a clear similarity between the oldest cratered terrain on Mercury and the lunar highlands, showing that they were cratered by the same population of impactors.

$N = D^{-3}$, where N is the number of craters per square kilometer within the size increment at diameter D. The power -3 has no fundamental physical significance, but it is mathematically convenient. We will be referring to relative size–frequency distribution plots in discussing the history of the icy moons. The concept may seem complicated, but to use these plots just two points need to be appreciated. First, the higher the value of R at a given crater diameter, the greater the density of craters. Second, if the crater distribution falls on a horizontal line then it has a straightforward impact history, but if the line is curved then there has either been more than one population of impactors involved or something has removed craters in part of the size-range subsequent to their formation. On a curved line, a segment that slopes upwards from left to right indicates relatively more larger craters, whereas a segment that slopes downwards indicates relatively fewer larger craters.

Figure 5.3 makes it easy to see that there are actually important differences in the crater distributions between the lunar highlands and the maria. Comparable differences in older and younger cratered areas have been found on Mercury and Mars. It is fairly clear that the post-maria craters represent impacts by asteroids and comets of similar characteristics to the bodies observed today still in orbit. The highlands bear both pre- and post-maria craters. Unless the subtly different crater distribution on the highlands is an artefact of some crater-modification process, it must indicate a somewhat different size–frequency distribution among the pre-maria impacting bodies at the time of the late heavy bombardment.

5.1.2 Crater formation

The foregoing discussion has made the implicit assumption that the craters we see on other worlds are the result of impacts and not of the other process capable of producing craters, namely volcanism. The origin of lunar craters was a hot topic for the hundred years or so leading up to the *Apollo* landings, with the proponents of impacts and of volcanism evenly matched. Nowadays the impact theory is very much the established view, although there are still a few heretics around.

One reason for supporting the impact origin of craters is the closely comparable size–frequency distributions on the Moon, Mercury, and Mars, which indicate a common external control. Moreover, a build-up of impact craters on an old surface seems inevitable, based on the measured current rate of impacts on the Earth's surface; for every 10 million km^2 (an area the size of Europe or the USA), a crater larger than 1 km in diameter is formed, on average, once every quarter of a million years, and one larger than 10 km in diameter every ten to twenty million years. Furthermore, the morphology of lunar craters is distinctly un-volcanic (for example, their floors are lower than the level of the surrounding terrain), whereas they have all the characteristics of natural impact craters on the Earth, of which many have now been studied—the famous Barringer meteor crater in Arizona, USA being perhaps the best known (Fig. 5.4)—and of impact craters formed in scaled-down laboratory experiments. This is not to say that there are no volcanic craters on the Moon, but they are far fewer than the impact craters and can be recognized by associated volcanic flow features and other characteristics.

Prototype for the Study of all Impact Craters in our Galaxy

Fig. 5.4. The Barringer meteor crater, Arizona, USA. (a) View of the 1.2 km diameter raised rim, from the north-west. (b) A rather exaggerated claim displayed on the approach to the crater, which at least serves to make the point that the processes by which extra-terrestrial impact craters formed are in little doubt, as a result of detailed field and laboratory studies on Earth.

In this book the conventional view will be followed that heavily cratered terrains are the result of a long history of impacts, and that a volcanic crater can (almost always) be recognized on morphological grounds. The processes that occur in an impact are by now fairly well understood. Typical impact velocities in the solar system are of the order of 10 km s^{-1}. At such high velocities, the first process that occurs when a projectile hits the target surface is that shock waves propagate forwards into the target and backwards into the projectile. For a fraction of a second the pressures are extremely high, far greater than those necessary to fracture the material of which both the target and the projectile are made, and indeed great enough to create high-pressure shocked forms of certain minerals, as has been documented both on the Earth and on the Moon. This short interval is known as the compression stage of an impact, and during it an upwards and outwards motion of ejecta consisting of vaporized, melted, and comminuted material from both the projectile and the target forms the beginning of an ejecta curtain (Fig. 5.5). By the time the projectile has completely disintegrated, the shocked region extends up to about one projectile diameter into the target.

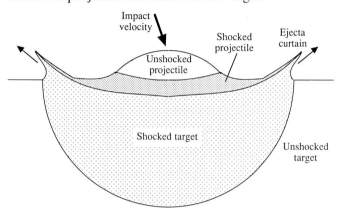

Fig. 5.5. The compression stage of an impact, a fraction of a second after arrival of the projectile, seen in cross-section. Melted and fragmented material derived from the shocked regions of the projectile and target feeds the beginning of a conical ejecta curtain. Material fed into the ejecta curtain at this stage has the highest velocities and will be the last and furthest to be deposited. The geometry is more or less independent of the angle at which the projectile hits the target surface.

The second stage of the crater-forming process, the excavation stage, continues with the growth of a transient cavity, feeding ever more material into the ejecta curtain (Fig. 5.6a). The transient cavity is hemispherical at first, but in a large impact it reaches a critical depth, after which it grows laterally only. It continues to grow until the energy of the impact is used up, and the last material to be supplied to the ejecta curtain joins it at low velocities, barely managing to flop out of the cavity, to form an overturned flap at the crater rim (Fig. 5.6b). The crater dimensions at the end of this stage are one to two orders of magnitude greater than those of the original projectile, and it takes about 10 s to form a 1 km diameter crater and about 100 s to form a 100 km diameter crater. The crater that remains may have the shape of the transient cavity as it was when the excavation stage petered out, but for craters larger than a threshold size (which depends on gravity and the composition of the target), inward collapse of the walls further enlarges the crater, with the formation of terraces and concentric rings. In addition, a central peak may form within in the crater, either as a result of ballistic rebound or due to push from the slumping walls (Fig. 5.6c).

On the Moon there is a progressive change in crater morphology as the size increases. Craters less than 10 km in diameter have simple bowl-shapes, whereas essentially all craters larger than about 40 km in diameter have flat floors, central peaks and terracing on the inner side of their rims. For those craters whose diameters exceed about 300 km it is common to find concentric rings dominating the crater, giving rise to the name 'multi-ringed basin'. The largest of these on the Moon have diameters of over 1000 km.

There are similar size-related morphologic transitions in craters on the icy moons. The critical diameters vary from one to another mainly because of their different surface

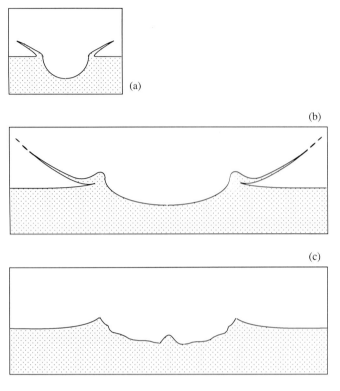

Fig. 5.6. Crater formation, seen in cross-section. (a) and (b) are during the excavation stage and (c) is after modification. In (a) the transient cavity is hemispherical. A conical ejecta curtain, consisting of material from both the projectile (now completely destroyed) and the target propagates outwards as the transient cavity grows. Stage (b) is several seconds or tens of seconds after (a) and is shown at about half the scale; the last material is entering the base of the ejecta curtain at low velocities and the transient cavity has reached its maximum extent. The upper, outer part of the ejecta curtain continues to spread and fall-out ballistically and will give rise to an ejecta blanket around the crater, including secondary craters where large blocks of ejecta hit the surface. During outwards movement, the ejecta curtain becomes increasingly separated into filaments that produce rays and loops of ejecta. In (c) the crater has been enlarged by inwards collapse of its walls and a central peak has formed.

gravities, and there is an additional class, central pit craters, which are not found on rocky worlds, except rarely on Mars where they may be related to water- or ice-saturated rock. The replacement of the central peak by a pit may be due to the gravitational collapse of a warm, possibly liquid, transient peak. The morphologic series with increasing scale on icy moons may be summarized as: bowl-shaped craters < central peak craters < central pit craters < multiringed basins. Examples of these are shown in Fig. 5.7.

5.1.3 Regolith

A surface that has been subjected to a prolonged period of bombardment loses its original physical constitution. In the case of the Moon, the surface is covered by a 'soil' or regolith of fragmented lunar rock and glassy material produced by impact melting. Less than about one per cent of the material present is derived directly from meteorites. The grain size of most of the regolith is less than about a tenth of a millimetre (although there are plenty of larger lumps), and this is thought to result from continual reworking of the surface by micrometeorite impacts as well as by the less frequent stirring up that occurs when a larger impactor strikes.

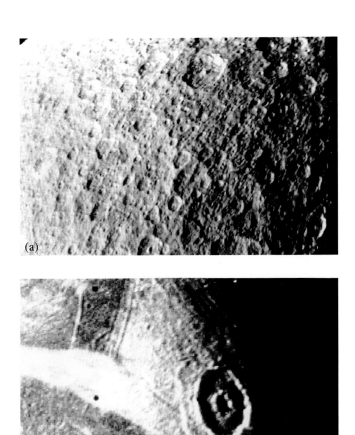

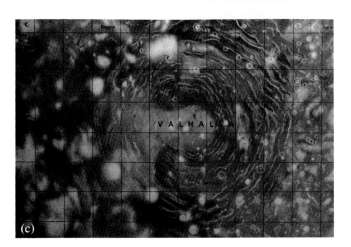

Fig. 5.7. The main crater morphologies that occur on icy moons with increasing scale. (a) Smaller bowl-shaped craters and larger central peak craters on Rhea. (b) A central pit crater on Ganymede. (c) An airbrush map showing the 4000 km diameter Valhalla multiringed basin on Callisto.

It is logical to expect that the cratered icy moons have similar regolith-covered surfaces. This is supported by three kinds of observation. The first is a study of the way the moons scatter sunlight as a function of incidence angle, which shows that their surfaces are granular. The second is Earth-based polarizing radar observations, which generally suggest a deep regolith of many tens of metres. The third is thermal infra-red measurements of the rates of cooling during eclipses, which demonstrate a low thermal conductivity compatible with a fragmentary surface layer. It is worthy of note that micrometeorite impacts probably help to concentrate rock dust in the regolith by ejecting water molecules or oxygen and hydrogen ions from the ice, in a process known as sputtering, which over long time periods results in a net loss of ice to space.

5.2 Impact cratering in the outer solar system

Cratering in the inner solar system is attributed to meteorites that are asteroidal fragments, with a proportion of cometary material. Jupiter lies beyond the main asteroid belt, and it is doubtful whether the supply of meteoritic debris is now or ever has been comparable with that in the inner solar system. The likelihood of the same population of impactors extending to Saturn and beyond is even less. Arguments about variations in the flux of crater-forming impactors throughout the solar system were at least partly resolved by the *Voyager* missions, which provided the first images on which craters could be counted and from which crater size–frequency plots could be drawn. Figure 5.8 shows a relative size–frequency distribution plot, like Fig. 5.3, with curves for Callisto and several more remote satellites. Without going into details just yet, neither the shapes nor the positions of the peaks of the curves for these bodies match those in the inner solar system. This implies that

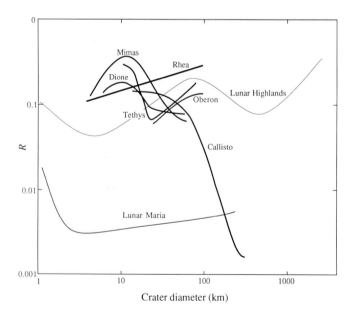

Fig. 5.8. Relative size–frequency distribution plot showing the difference in crater populations between the Moon and the most densely cratered regions of several heavily cratered outer planet satellites.

the size–frequency distribution of the impacting bodies themselves must have been different. There are also significant differences in the curves between Callisto and the moons of Saturn, and even within the family of Saturn's moons.

5.2.1 Crater scaling between satellites

When comparing crater statistics between diverse bodies, it is well to be aware of the pitfalls. An impactor with a given mass and impact velocity will produce a smaller crater on a world with high surface gravity than on one with lower surface gravity. In addition, target material such as ice that has a comparatively low surface density will form larger craters than a rocky target with high surface density. The predicted values depend on the details of the model used, but in general crater diameters on the icy satellites should be about one and a half to three times greater than on the Moon, for the same impact conditions.

The sizes at which craters undergo transitions in morphology also vary, and this is something we can observe directly. The transition from simple bowl shapes to more complex craters with a central peak occurs below 10 km for Callisto and Ganymede, but at around 30 km for smaller worlds such as Mimas. Despite their slightly lower surface gravity than the Moon, the morphologic transition on Callisto and Ganymede occurs at a smaller size than it does there, which is evidently a result of the lower strength of their icy surface material than the Moon's rocky surface.

There are also considerations involving the orbital motion of a synchronously rotating satellite and the gravity field of its planet. Impacts with projectiles originating beyond the satellite system in question should be most common on the leading hemisphere of a satellite, because this is the side that is travelling towards danger. In the case of a satellite in synchronous rotation, one hemisphere is fixed in this role, so it should accumulate several times more craters than the trailing hemisphere if the tidal lock has remained unbroken throughout the satellite's impact history. The effect increases with the orbital velocity of the satellite, so it should be more pronounced on the inner satellites of a planet, which travel the fastest. Externally originating projectiles are subject to another important effect in that incoming projectiles will be deflected towards the planet by its gravitational field, so they are more likely to hit an inner satellite than an outer one. The effect of this gravitational focusing mechanism would be to make the present cratering rate on Mimas (Saturn's innermost major satellite) about twenty times that on Iapetus (its outermost big satellite).

5.2.2 Impactor populations

Scaling effects therefore appear to be capable of shifting the crater size–frequency distributions for icy moons towards larger diameters by up to a factor of about three, and up and down by an order of magnitude or so. There is no way that this could account for the disparities between the distributions shown in Fig. 5.8, where there are clearly different slopes. This problem lends support to the concept that different populations of impactors have been active in each part of the solar system.

The most obvious trait in the crater size–frequency distribution for Callisto is a drop-off at diameters greater than about 60 km. This is repeated in comparable data for Ganymede (not shown in Fig. 5.8). It appears, then, that the population of impactors affecting Jupiter's satellites was deficient in impactors sufficiently massive to create large craters. As we shall see shortly, some large craters have probably been lost because they subsided under their own weight when the lithosphere was thin and warm, but it is unrealistic to appeal to this mechanism to account entirely for the loss of the large craters. Thus most people agree that Callisto and Ganymede record a flux of impactors substantially different to that affecting the inner solar system. A likely explanation is that the outer population was (and is) dominated by comets, whereas the population that has affected the inner solar system since the late heavy bombardment is dominated by asteroidal material. An unfortunate corollary of this theory is that we cannot use crater densities at Jupiter for absolute dating, because there are no means of linking these crater statistics to independently determined ages from the Moon and the Earth.

At Saturn, the situation is even less satisfactory. There are considerable differences in the crater size–frequency distributions even among its heavily cratered moons. No moon of Saturn shows a correlation between the density of craters and position on the globe. This suggests that the tidal-locking of the rotation was broken several times by impacts large enough to impart sufficient angular momentum to increase the rate of spin temporarily, and that each time synchronous rotation was restored (by tidal forces) the moon presented a different face to its planet. Within the Saturn system as a whole, two populations of impactors have been proposed, known as population I and population II. Population I is older, and its traces have been partly erased on younger surfaces on Dione and Tethys. It produced a substantial proportion of large craters. Population II yielded a deficiency of large craters and is the dominant population on younger surfaces. On Fig. 5.8 the curves for Rhea and the upturn in the curve for Tethys are due to population I, whereas the peaks at about 10 km diameter for Mimas, Dione, and Tethys are due to population II. A similar pair of populations is recognized on the moons of Uranus, the curve for Oberon being essentially a population I trace.

A reasonable explanation for the population I craters is that they were formed by sweeping up post-accretional debris, possibly ending around 4 billion years ago, and are more or less equivalent to the late heavy bombardment that has been documented in the inner solar system. Their size–frequency distribution is not entirely dissimilar to that of the lunar highlands, the degree of difference being quite reasonable in view of their vast separation in space, one result of which would have been to give a preponderance of icy rather than rocky material in the impacting population. Population II craters represent a discrete episode of cratering, with a size–frequency distribution that cannot be reconciled with anything seen at Jupiter. They are usually attributed to collisions with a swarm of debris in orbit about the planet. Such a debris swarm could have been created by a single major impact within the satellite system, one that completely destroyed a satellite or at least knocked large chunks out of one. It follows that there is no reason why the population II impactors at Saturn and Uranus should have been around at the same time as each other.

If this explanation for populations I and II is correct, then neither is related to the cratering by cometary impact that must continue to the present day (sometimes known as population III), and which must have carried on at its background level while the population II flurry or flurries of activity occurred. Unfortunately, the size–frequency distribution at Jupiter (as expressed by Callisto and Ganymede) is at its most distinctive for crater diameters greater than those possible on the moons of Saturn and Uranus (which are much smaller worlds), so it is difficult to tell how the impactors that caused them compare with population I. It is possible that the crater population observed on Callisto obliterates the traces of an original epoch of population I cratering. Thus, relative dating between one satellite system and another, and between the inner and outer solar system, is fraught with caveats and uncertainties. The sanguine view is that the problem may never be solved, even when the badly needed higher resolution images are available to supply the data that are often lacking in the 1–20 km range, but at least craters give us some idea of the external influences that have moulded the shapes of many of the worlds in the solar system.

Enough of generalities. At last we are in a position to examine individual worlds in detail. We will begin with Callisto, the largest icy moon in the solar system to fall into the heavily cratered, 'dead worlds' class.

5.3 Callisto

Callisto is the second largest and least dense of the galilean satellites (Table 1.1). Extreme models for its differentiation were shown in Fig. 2.5. The images sent back by the *Voyager* probes revealed Callisto as an intensely cratered world (Fig. 5.9), although with a dearth of craters in excess of about 60 km diameter. Its surface is one of the darkest known among the icy satellites, with an albedo of about 0.2. If we accept the great age of Callisto's surface implied by its crater density, then the darkening can be explained by the presence of a residual 'lag deposit' rich in rock dust left on the surface of the regolith as ice has selectively been removed by vaporization during impact and by sublimation. The younger craters and the rays and other ejecta emanating from them have higher albedos, around 0.4, that fade into the background with increasing age (Fig. 5.10). This indicates the penetration of impacts into a more ice-rich (but still impure) subsurface and dispersal of this fresher material in the ejecta. However, the presence of a significant proportion of rocky material near the surface shows that any differentiation that has occurred in Callisto did not extend to its outer layers.

5.3.1 *Multiringed basins*

In addition to its craters, Callisto has several multiringed structures, the four largest of which are located in its leading hemisphere, which is more likely to suffer impacts by debris not in orbit around Jupiter. The second largest of these, named Asgard (in keeping with the far-northern name convention chosen for Callisto), shows up rather poorly in the

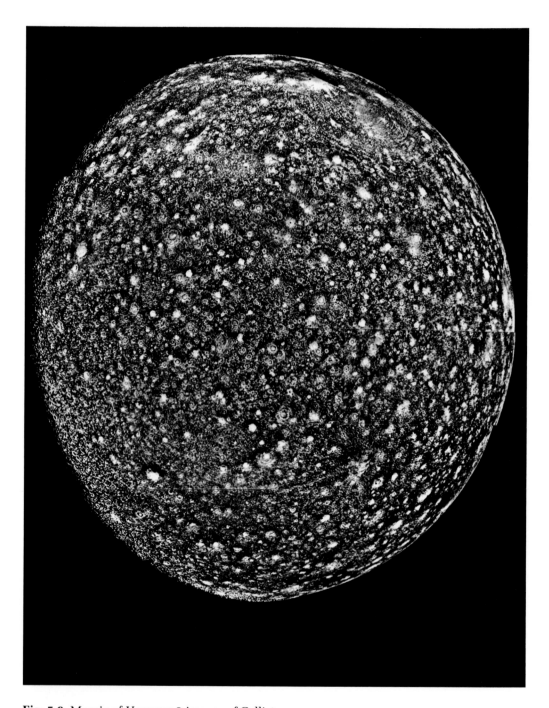

Fig. 5.9. Mosaic of *Voyager-2* images of Callisto.

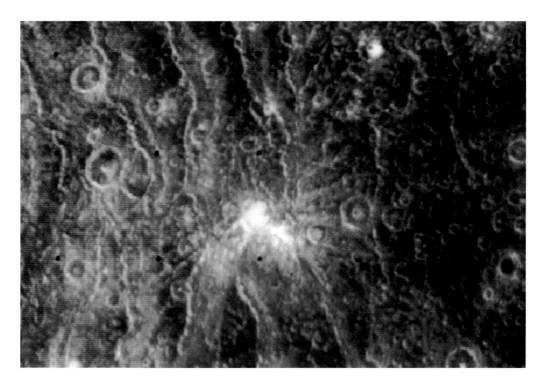

Fig. 5.10. Craters on Callisto. The youngest craters are bright and progressively older craters fade into the background. The largest craters here (which have central pits) are about 60 km in diameter.

top right of Fig. 5.9, and Fig. 5.11 shows the largest structure, Valhalla, somewhat more clearly. The outermost ring of Valhalla is some 2000 km from its centre, giving it a diameter of about 4000 km. This makes it by far the largest multiringed basin in the solar system; in comparison, the Procellarum, Imbrium, and Orientale (Fig. 5.12) basins on the Moon are 3200, 1500, and 930 km in diameter, respectively, whereas the Caloris basin on Mercury reaches only 1300 km in diameter.

Despite the similarity of the multiringed basins on Callisto to those on the Moon and Mercury, there are many subtle differences that serve as clues to the strength and thickness of Callisto's lithosphere at the time of their formation. In the centre of each of Valhalla and Asgard there is a bright region, 600 and 230 km in diameter, respectively, that is taken to be the remains of a central structure which collapsed under its own weight, either because it was warm and mushy when it formed or because the lithosphere was too thin to support its weight. The central part of the pale area is particularly flat. To describe such a feature as this, which has lost its original morphology and is identified essentially by its anomalous brightness and smoothness, planetary scientists have appropriated the term 'palimpsest', which originally meant an ancient parchment on which the original text had been erased and that had been subsequently overwritten. The Valhalla palimpsest is

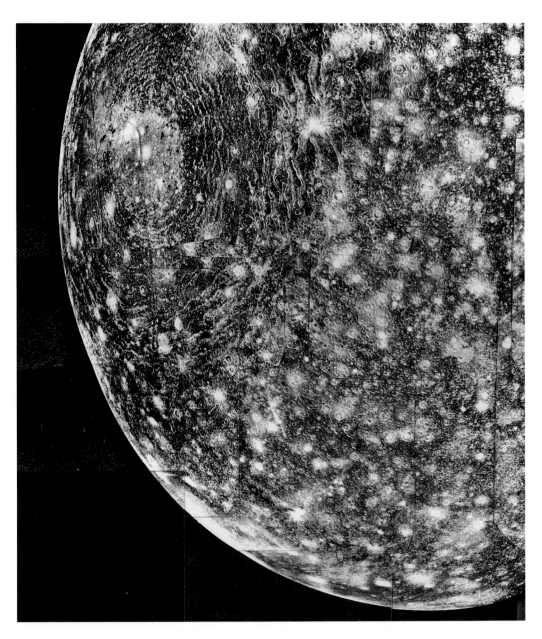

Fig. 5.11. Mosaic of *Voyager-1* images, showing most of the 4000 km diameter Valhalla basin on Callisto. Compare with the airbrush map of this feature in Fig. 5.7(c).

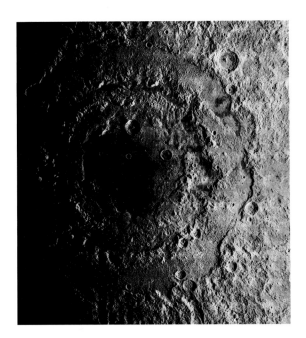

Fig. 5.12. Mosaic of *Lunar Orbiter* images, showing the 930 km diameter Orientale basin on the Moon. Compare with the Valhalla basin on Callisto in Fig. 5.11.

particularly well seen on Fig. 5.11. The relative ages of the multiringed basins are given by their relationships to smaller craters; for both Asgard and Valhalla the central palimpsests have about one-third of the density of superimposed craters seen on the surrounding terrain, and of those craters that intersect a ring about one-third are clearly younger than the ring, whereas the remainder are overprinted by it.

Although the palimpsests in the centres of Callisto's basins contain no direct evidence of their origin, there is plenty that can be inferred from the concentric rings around each basin. In the case of Valhalla the rings are about 15 km wide and are 20–30 km apart in the inner part of the basin, rising to 50–100 km apart in the outer region. No single ring runs right round the structure; instead they have the shape of sinuous arcs that are mostly between 200 and 500 km in length. For a distance of 200–300 km outward from the bright central region the rings consist of ridges less than 1 km in height that are paler than the terrain in which they lie. Beyond this the rings are dominantly outward-facing scarp slopes 1–2 km high, although a few are inward-facing scarps and some near the inner edge of this outer zone even take the form of troughs. Where a ridge cuts a pre-existing crater, the crater remains circular with no measurable shortening or stretching. Similarly when a trough cuts through a crater, the portion of the crater within the trough appears to have been simply down-dropped, with no change in its overall shape, in a structure that a geologist would call a graben. A pale hummocky unit hugs the base of many of the scarps and probably represents material extruded as a melt or as a crystal-rich icy mush along these fractures. The same unit appears to have inundated the down-dropped portions of some of the craters within the troughs. Several of the outward-facing scarps in the Valhalla ring system are visible running from the top to the bottom of Fig. 5.10.

There have been several attempts to explain the morphology of the Valhalla basin in terms of the response of Callisto's lithosphere and asthenosphere to the formation of a large transient cavity caused by a major impact. Essentially, the thinner the lithosphere the more rings are expected, and the closer their spacing. The rarity of radial features in Valhalla provides additional clues to strengths and thicknesses. The details need not bother us here, but it appears that Valhalla is best explained by an impact into a lithosphere only about 30 km thick, the strength of which was intermediate between that of the Ross Ice Shelf (Antarctica) and that of the Earth's lithosphere. A liquid asthenosphere can be ruled out because that would have resulted in a very different morphology, and it seems clear that the asthenosphere deformed by solid-state flow. A schematic model is shown in Fig. 5.13; in response to the formation of a transient cavity the nearby asthenosphere flowed inward to fill the void. This exerted an inward drag on the base of the lithosphere, causing it to fragment into concentric zones. At the same time, subsidence of the raised rim drove a weak asthenospheric flow away from the crater. The inward-flowing region gave rise to the inward-facing scarps and occasional grabens, and the ridges closer to the centre are interpreted as extrusions derived through concentric fractures in the lithosphere. The surrounding, weaker, outward flow produced the outward-facing scarps.

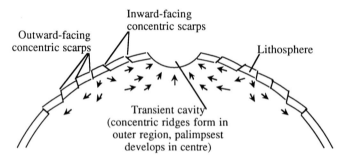

Fig. 5.13. Model for the generation of multiringed basins on Callisto, showing the asthenospheric flow induced in response to the excavation of a transient cavity by a major impact, and the concentric fractures that this flow causes in the overlying lithosphere.

One complication to the generalizations about the thickness of an icy lithosphere that were made in Chapter 4 is that the deformation behaviour of ice depends not just on the imposed rate of deformation (the *strain*) but also on the magnitude of the deforming force, that is on the *stress* (in particular the shear stress). The stresses caused by the presence of a transient cavity are so high that the lower part of what would otherwise be called the lithosphere behaves as part of the asthenosphere. The 30 km thickness quoted for Callisto's lithosphere at the time of the formation of Valhalla is thus applicable to the impact-defined lithosphere only. Internally generated (geological) stresses would almost certainly be less, and the effective lithosphere for all but the virtually instantaneous excavation of deep cavities by impacts would be greater than this by a factor of at least two or three times.

5.3.2 Crater relaxation and retention

Callisto contains abundant evidence of the cooling and thickening of its lithosphere (however defined) through time. The topography of most craters in all size ranges is subdued, and the effect is greatest for the oldest and largest craters. This is presumed to have occurred by a process known as viscous relaxation in which the lithospheric ice deformed very slowly under its own weight, tending to even out the variations in topography (Fig. 5.14). To do this it must have been warmer than it is today, and probably thinner as well. Even so, the time required to virtually wipe out the topography of a crater 20 km in diameter could have been of the order of a billion years. The viscous relaxation of Callisto's early large craters so that they disappeared completely is probably a contributory factor to the deficiency of larger craters in its crater size–frequency distribution.

Original topography

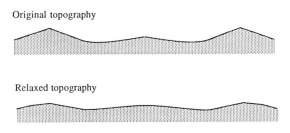

Relaxed topography

Fig. 5.14. Relaxation of crater topography, such as occurs when a relatively large crater is formed in a warm, thin lithosphere. This effect can be seen in the form of many of Callisto's craters.

The relaxation of craters takes on a special significance when trying to interpret Callisto's impact history and lithospheric evolution. The four largest multiringed structures are in the leading hemispheres, and craters more than 60 km in diameter are twice as abundant near its apex of orbital motion (i.e. in the middle of its leading hemisphere) than near the antapex of motion, but there is no leading–trailing asymmetry in the density of smaller craters. If the impacting bodies originated externally to the Jupiter system, nearly ten times as many craters would be expected at the apex of orbital motion as at the antapex. One way to explain this is that most of the impactors (and all of the smaller ones) originated *within* the Jupiter system; an alternative model suggests that there was a leading–trailing asymmetry in cratering but that most of the evidence has been lost as a result of asymmetric strengthening of Callisto's lithosphere over time. In a strengthening lithosphere the oldest craters to survive should be the smaller ones (<10 km diameter), and the proponents of the asymmetric cratering model for Callisto claim that the oldest survivors of this size occur around the antapex of its orbital motion. This implies that the lithosphere cooled and thickened faster there than in the leading hemisphere, where the record of such ancient craters has been lost due to viscous relaxation.

Slower cooling and thickening in Callisto's leading hemisphere can be explained if it was blanketed by a thicker layer of regolith, the thermally insulating properties of which would keep the local lithosphere warmer and thinner. A thicker regolith on the leading hemisphere is exactly what would be expected as the result of a greater number of impacts in that hemisphere, by impactors derived from beyond the Jupiter system. Similar asymmetries in the supposed crater retention ages and crater densities on Ganymede support this view. If true, this model implies that Callisto was in synchronous rotation from a very early time (which would not be surprising in view of the tidal despinning model discussed in Chapter 2) and that the lithosphere was fixed in place with respect to the globe as a whole, unable to slide around freely on top of the asthenosphere.

The lithosphere of Callisto appears by now to have grown too thick to allow the formation of multiringed basins even in response to the biggest impacts; this is shown by the presence of one bright, fresh, crater (Adlinda) with a diameter of about 150 km and with no sign of rings around it like those that were produced by inward collapse into Valhalla, Asgard, and the other older basins.

Setting aside the unresolved arguments about the source of the impacting bodies and the degree of leading–trailing asymmetry in cratering, we can summarize the evolution of Callisto, as indicated by the nature of its cratering record, as follows. After accretion and whatever degree of differentiation occurred, the lithosphere was at first too warm to retain any recognizable craters. This elevated temperature was due to a combination of residual heat from accretion and heat generated by tidal despinning, differentiation and radio-activity in Callisto's rocky component. With time, as the heat from the first three sources leaked away and the rate of radiogenic heat production declined, the lithosphere cooled and thickened sufficiently to allow the retention of smaller and then progressively larger craters. As will be seen in the next chapter, this is a story repeated at Ganymede, but with the addition that Ganymede has experienced major tectonic resurfacing.

5.3.3 *Frost migration*

Callisto today is essentially an inert world. The next time a 10 km diameter crater forms on Callisto it will just sit there, without relaxing, indefinitely (until it is obliterated by subsequent craters) because the lithosphere is now very cold and thick.

There are towns on Earth where they say the most exciting thing to do on a Saturday night is to go out and watch the traffic lights change. The equivalent on Callisto would be to go and watch the frost migrating. Figure 5.15 shows a portion of Callisto's surface at high northern latitudes. The inside walls of most of the craters appear brightest on the south side of the crater. This is clearly not an illumination effect because these slopes face away from the Sun, and it is the opposite slopes, on the south-facing inside wall on the north side of each crater, that are the most fully illuminated.

The mid-day temperature on Callisto is about 116 K at 60° N. This is sufficient to allow a significant amount of sublimation of ice (i.e. molecules escaping directly from the solid to the vapour phase) over geological time. It has been calculated that a south-facing slope of a crater at 60° N would lose ice by sublimation at a rate of about 60 m per billion years, until the process became checked by the concentration of a rocky residue over the surface. In contrast, sublimation from a north-facing slope, where the temperature is probably some 30 K lower, would be at less than a thousandth of this rate. The most likely explanation of the whitening of north-facing crater walls is that ice has sublimed in the more fully sunlit and therefore warmer areas nearby and has been preferentially redeposited on the colder, north-facing slopes.

We will now turn to the other heavily cratered 'dead worlds' out at Saturn and Uranus, where the surface temperatures are too low to allow even the limited relief to monotony provided by sublimation and frost migration. As we have already seen, their differing crater size–frequency distributions are indicative of at least two populations of impactors.

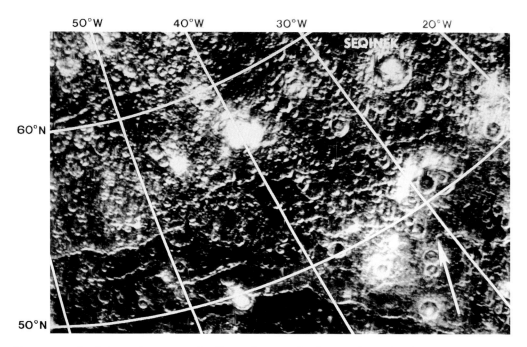

Fig. 5.15. A high-latitude region of Callisto showing the frost-induced brightening of the north-facing crater walls. This is particularly well shown within the 60 km diameter crater Seqinek, towards the upper right of the picture. The illumination direction is shown by an arrow.

Aside from that, the main interest in these other heavily cratered worlds lies in how and why they differ from their more recently active neighbours.

5.4 Rhea

Rhea lies in the middle of Saturn's family of moons (Fig. 1.2). With a radius of 764 km it is considerably smaller than Callisto, but is the largest of the other dead worlds. The bulk density measurements made during the *Voyager* fly-bys suggest that the composition of Rhea is about 40 per cent rock if it is a differentiated world, or about 35 per cent rock if it is undifferentiated, the remainder being water-ice with probably a small fraction of methane and ammonia. Its albedo is high, about 0.6, suggesting that it has a much cleaner icy surface than Callisto. Most of its Saturn-facing hemisphere was imaged at a resolution of 20 km, increasing to better than 2 km in the north, but the anti-Saturn hemisphere was much more poorly covered (resolution about 50 km) so we know much less about it. In the well-imaged areas, Rhea appears very much the archetypal heavily cratered icy satellite. In general, large craters of all ages show no evidence of viscous relaxation of their morphology (Fig. 5.16, Fig. 1.4), offering evidence of more rapid cooling and thickening of the lithosphere and also of the lower gravity compared to Callisto.

Fig. 5.16. *Voyager-1* image of part of Rhea's north polar region, about 300 km across. The craters show no evidence of viscous relaxation of their topography. The polygonal outline of some of the craters may be due to impact into a deep rubble layer. Very bright slopes on some inner walls of the craters may be relatively clean ice exposed by slumping.

As shown in Fig. 5.8, the most densely cratered region of Rhea is dominated by population I craters, with a positive slope on the relative size–frequency distribution plot. However, there are regions such as near Rhea's north pole where population II craters predominate, suggesting a resurfacing event towards the end of the population I bombardment that some workers have attributed to mantling by ejecta from two poorly-imaged large (possibly multiringed) basins and others to episodes of some kind of endogenic resurfacing. Possible clues that there was endogenic activity on Rhea are given by a few linear troughs that run through population I terrain but that cannot be traced into younger population II areas; Kun Lun Chasma, shown in Fig. 3.8, is an example. These could record an episode of thermal expansion as Rhea heated up, or a volume change associated with differentiation (although they seem to be too young for both of these), but could equally well be impact-generated fractures.

The general lack of evidence of geological activity on Rhea is surprising in view of its probable thermal history since formation. Two models for Rhea's internal temperature evolution over time are shown in Fig. 5.17, differing essentially in how much accretional heat is retained. Warming beyond the initial state is modelled on the basis of radiogenic heating, using the ice to rock ratio derived from the bulk density measurements of the

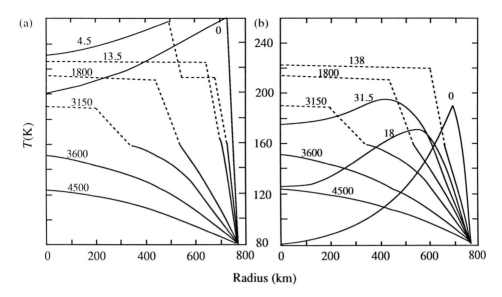

Fig. 5.17. Two models of the thermal evolution of Rhea. (a) 'Hot start', with a large degree of accretional heat retained and (b) 'cold start'. These plots show the internal temperature profiles at various times, in millions of years after formation. The solid lines indicate heat transported by conduction, the broken lines show heat transferred by solid-state convection. The models are very similar from about 100 million years after formation and show a convective asthenosphere enduring until about 3.3 billion years, below an ever-thickening lithosphere.

Voyager experiments. On both models the interior warmed up (although never to melting point) and began to convect in the solid state about a few tens of millions of years after formation, and continued to convect to within 1.5 billion years of the present day, by which time the whole globe had become effectively lithospheric. The freezing of any traces of liquid ammonia hydrate would have been complete at about the same time. Presumably subsequent cratering has obliterated traces of surface tectonism (with the dubious exception of the linear troughs) that occurred while the lithosphere was still thin enough to be fractured and deformed by internal convection, but this makes it difficult to explain the clear signs of tectonism on many of Saturn's smaller and generally less dense satellites that we will encounter in the next chapter.

Another puzzle posed by Rhea is why there are no signs of any global contraction in volume. As noted in Chapter 2, Rhea is large enough for ice II to be the stable phase near its centre. This is the case whether we accept an undifferentiated (Fig. 5.18a) or a differentiated (Fig. 5.18b) model for Rhea's internal structure. Referring back to the ice phase diagram in Fig. 2.4, it can be seen that the ice II–ice I phase transition shifts towards lower pressures as the temperature decreases. This means that as Rhea cooled, the central ice II zone would have grown at the expense of the overlying ice I zone. Ice II is about 20 per cent denser than ice I, and the conversion of ice I to ice II as the temperature fell should have resulted in a global contraction of Rhea's radius by about 15 km, leading to a

(a) 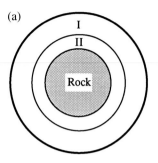   (b)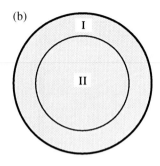

Fig. 5.18. Two extreme models for the internal composition of Rhea, showing the stability fields of ice I and ice II. (a) Fully differentiated, with a rocky core; and (b) undifferentiated.

reduction in its surface area of 4 per cent. This should have occurred gradually during the later part of Rhea's history, but it is difficult to reconcile the observed linear troughs with compression.

To have a better chance of understanding Rhea we need higher resolution images of the anti-Saturn hemisphere, the trailing part of which is darker than the rest of the satellite, but which has pale wisps superimposed on it that could be the missing tectonic features. With the 50 km resolution images available at present, there is little that can be said other than that parts of Rhea evidently underwent some kind of resurfacing event towards the end of population I time (which may represent the end of the late heavy bombardment), but we do not know what form this resurfacing took. For all we know, in the poorly-imaged hemisphere there could be vast tracts of internally reheated and remobilized terrain, or volcanic vents that could have obliterated some of the population I terrain with a mantle of 'pyroclastic' ice. Such features would make it more straightforward to interpret the rather uninformative hemisphere of which we do have good images. For the time being, Rhea must be consigned to the 'dead worlds' class, with the remark that in the next chapter we shall see examples of worlds where past episodes of resurfacing are much more obvious.

5.5 Iapetus

Iapetus, the outermost of Saturn's major moons, is a strange world. In terms of size and density it is a close twin of Rhea (Table 1.1) and would be expected to have a similar internal structure and thermal history. Long before the space age Iapetus was famous for the dramatic difference in albedo between its leading and trailing hemispheres. The leading hemisphere has an overall albedo of only 0.1, whereas that of the trailing hemisphere is around 0.5, a value typical among icy satellites. This means that, through a telescope, Iapetus appears considerably darker when to the east of Saturn in the sky (when its leading hemisphere is seen) than when it is to the west of the planet (when its trailing hemisphere is seen). This is just the sort of anomaly that science fiction writers like to get their teeth into; for example, in the book (although not the film) of *2001, A Space Odyssey*, the 'Star Gate' monolith erected by alien intelligences was located in the exact centre of the bright hemisphere.

Unfortunately, the *Voyager* probes did not find anything quite so amazing, although Iapetus remains something of an anomaly and an enigma (Fig. 5.19). Unluckily neither probe was able to pass very close to Iapetus; *Voyager-1* provided images of parts of the Saturn-facing hemisphere with a maximum resolution of about 50 km, and *Voyager-2* imaged primarily the anti-Saturn hemisphere rather better with a resolution that reached about 20 km in places. The best images revealed that the bright terrain is heavily cratered for diameters greater than 30 km. Smaller craters were too small to see on the images, but by extrapolating the size–frequency distribution to smaller diameters the total crater density would seem to be comparable with densely cratered objects such as Mercury, Callisto, Rhea, and the lunar highlands. There are no traces of any tectonic activity.

The dark hemisphere is so dark that the *Voyager* imaging system was not able to record any topographic details, such as craters, within it (although this might have been achieved with the kind of long exposures and motion-compensation used at Uranus and Neptune), so the nature and relative age of the dark material remains unclear. What we do know is that the distribution of the dark material is symmetric about the apex of orbital motion, with an albedo of only 0.02 at the apex, decreasing to 0.04 near the edges of the dark terrain. This sort of distribution argues strongly for an external control of the darkening process. Ground-based photometry shows that the dark material is very red, and it has been suggested that its composition can be explained by hydrated silicates with a small admixture of organic polymers of the sort known to occur in carbonaceous chondrite meteorites. A possible source for this material is impact-eroded dust from Phoebe, the next moon out from Saturn, which is a non-spherical body just over 200 km across that is in a retrograde orbit nearly four times further out than Iapetus. Phoebe has a similar albedo to the dark side of Iapetus, but it is less red, so dust from Phoebe may not be the sole component of the dark material on Iapetus. Alternative sources might be other, hitherto undiscovered, minor satellites in retrograde orbits beyond Iapetus, or carbonaceous dust from within Iapetus itself that has been preferentially accumulated on the surface of its leading hemisphere due to impact-reworking and volatilization of the surface ice.

If the darkening is due to a surface coating it must be either very young or very thick because there are no examples of craters penetrating the dark material and depositing paler ejecta on top of it, or indeed of pale spots of any kind within the dark hemisphere. A young age is also indicated by the asymmetry, because Iapetus, being so far from Saturn, is only weakly tidally locked, and large impacting projectiles of the sort that would have formed its largest craters probably carried enough momentum to temporarily change the rate of spin, with the result that the leading hemisphere would have changed its position on each occasion. Over time, this would lead to a roughly even distribution of dust across the surface of Iapetus, so the logical explanation is that the darkening we see is a relatively young phenomenon.

One puzzle that remains with any of the proposed externally controlled origins for the dark material is that, as shown on Fig. 5.19, there are craters at the edge of the pale hemisphere that have dark floors. Externally derived dark material should coat crater floors, crater walls and the surrounding terrain equally, and the presence of dark floors only would seem to indicate that at least some of the dark material may come from within

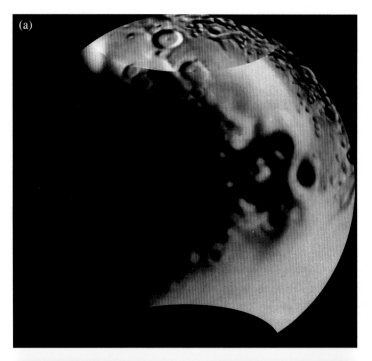

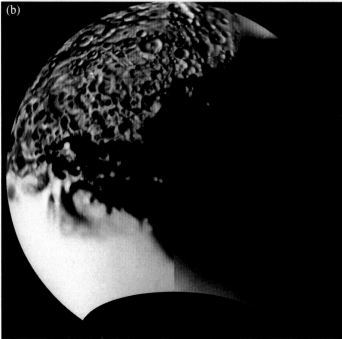

Fig. 5.19. Shaded relief maps of Iapetus, showing (a) the Saturn-facing hemisphere and (b) the anti-Saturn hemisphere. The leading hemisphere is too dark to allow any details within it to be visible on the *Voyager* images. The unimaged south polar region is omitted. These (and all the other shaded relief maps of complete hemispheres in this book) are equal area projections.

Iapetus. Unfortunately, there are no spectral data to confirm whether or not the dark crater floor material has the same composition as the dark hemisphere coating.

It remains a possibility that the dark material has been deposited by volcanic processes. However, if this were the case it would be a remarkable coincidence that the dark material is distributed so symmetrically about the apex of motion. Iapetus then is probably a 'dead moon', from a geological if not from a dust-gathering point of view.

5.6 Mimas

Mimas is the smallest and innermost of Saturn's regular satellites. Being so small, neither accretional energy nor radiometric heating is likely to have been sufficient to drive convection at any time in the past with sufficient vigour to have left traces on its surface, as indicated by the thermal models in Fig. 5.20.

Mimas has been imaged with a resolution as good as 2 km in much of its southern hemisphere, although the northern hemisphere coverage is markedly poorer. The surface is heavily, but not uniformly, cratered. Overall, the crater size–frequency distribution for Mimas is dominated by population II cratering (Fig. 5.8) and large tracts of the surface are devoid of craters greater than 30 km in diameter, suggesting that these areas have been resurfaced. Mimas' most striking feature (no pun intended) is a well-preserved 130 km diameter crater (Fig. 5.21), named Herschel, after Sir William Herschel who discovered Mimas in 1789. A few troughs crossing the surface roughly radially to the centre of Herschel are evidence of the severe effect of this particular impact, which was probably close to the maximum size that Mimas could have sustained without breaking apart. This is a salutary reminder that a satellite can be destroyed if a sufficiently energetic impact occurs. Most of the debris from such an event would subsequently re-accrete to form the satellite anew, but the new surface, and that of its neighbours, would experience a new interval of bombardment until the last fragments were swept up. If this has gone on, then

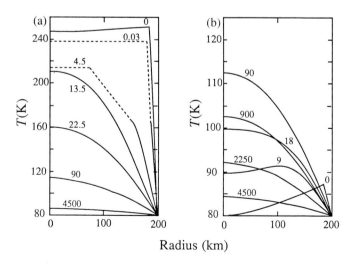

Fig. 5.20. Two models of the thermal evolution of Mimas. (a) 'Hot start' and (b) 'cold start'. These plots show the internal temperature profiles at various times, in millions of years, after formation in the same manner as Fig. 5.17. Mimas is unlikely ever to have experienced prolonged convection and is almost certainly undifferentiated.

Fig. 5.21. *Voyager-2* image of Mimas, showing the enormous crater Herschel, which is one-third the diameter of the moon itself and has a depth of about 10 km and a central peak about 6 km high.

we are chasing shadows when we try to construct time-scales based on crater statistics in the outer solar system. It is possible that any satellite we see today with a diameter less than about 1000 km has been broken up and re-assembled from its own fragments one or more times since its initial formation. Its subsequent differentiation would be much the same as the first time around, except that as time went by there would be progressively less radiogenic heat to assist with the process. As we shall see shortly, impact fragmentation and re-accretion is particularly likely to have happened to the inner satellites of Uranus.

5.7 Oberon

The two remaining 'dead worlds' are Oberon and Umbriel, the outermost and third-outermost of the satellites of Uranus. The bulk density measurements made of the five major satellites of Uranus during the *Voyager-2* fly-by suggested that there are no clear compositional differences between them. However, opinion is divided as to whether the carbon that must be present (unless the present models of solar system chemistry are very badly wrong) is dominantly organic or a mixture of methane and graphite. Some alternative chemical compositions are given in Table 5.1.

Despite their apparent uniformity in composition, the satellites of Uranus display a striking range of geological histories, even when comparing moons of a similar size; for example, Oberon has a different history to Titania, its twin in size and mass. However, present judgements should be treated with caution, because the orientation of the Uranus

Table 5.1 Two models for the mass fractions of the chemical constituents of the satellites of Uranus. The carbon in model 2 could occur in organic molecules instead of graphite; this would require the rock fraction to be greater, in order to be compatible with the observed bulk density

Chemical constituent	Model 1	Model 2
Silicate rock	0.336	0.298
Water-ice	0.511	0.455
Ammonia-ice	0.074	0.065
Methane clathrate	0.079	0.070
Graphite	0.0	0.112

system at the time of the *Voyager-2* fly-by meant that only the southern hemispheres of the moons were in sunlight, so we have images covering only half of each world. Furthermore, because of the bull's-eye geometry of the encounter (Fig. 3.7a) the resolution of the images is poor for the outer satellites, being about 20 km in the case of Oberon.

Voyager-2 revealed Oberon to be a heavily cratered world, within the limits imposed by the 20 km resolution of the best images (Fig. 5.22). The surface is generally fairly dark, with an albedo of about 0.25, but what appear to be the youngest of the larger craters are surrounded by brighter ejecta blankets and ejecta rays (Fig. 5.23). This could indicate a

Fig. 5.22. *Voyager-2* image of Oberon. Note the peak seen on the limb towards the upper left.

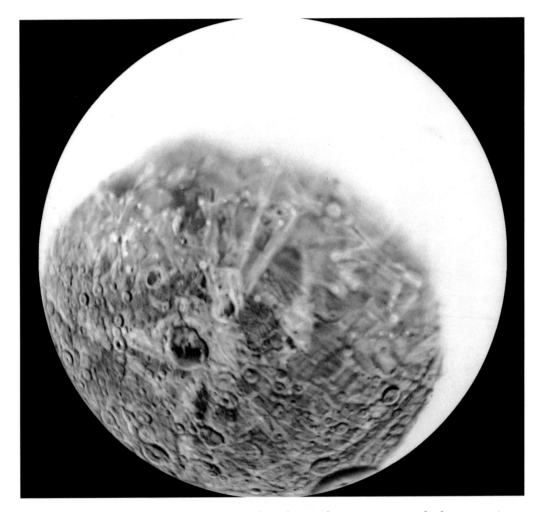

Fig. 5.23. Shaded relief map of Oberon's southern hemisphere, approximately the same orientation as Fig. 5.22. Note the bright ejecta associated with dark-floored craters. The unimaged area has been left blank.

dust-rich surface overlying an icier substrate, as suggested earlier to explain a similar phenomenon on Callisto. An alternative explanation is that what was originally methane incorporated in the surface ice (as a clathrate) has been damaged by exposure to cosmic radiation, in particular high-energy electrons trapped in Uranus' magnetosphere. It has been calculated that bonds between carbon and hydrogen would be broken as a result of this radiation at a rate sufficient to convert any methane in the upper millimetre of Oberon's surface to dark organic compounds and carbon in less than ten million years. This is an attractive idea, in that it explains both the general darkness of all the satellites of Uranus and why methane cannot be detected spectroscopically.

Various other characteristics of Oberon can be seen on Figs 5.22 and 5.23. One is the dark infill on some of the crater floors, reminiscent of Iapetus near the edge of its leading hemisphere. Another is a mountain peak some 20 km high seen in profile on the limb in Fig. 5.22. This could be the central peak of an impact structure, but whatever it is, its presence is proof of a thick, strong lithosphere. Enhanced images show vague hints of a few gently curved fault-like features running across the disc (represented in the lower part of Fig. 5.23), which could be a consequence either of impacts or of an episode of global tectonism. It is reasonable though to class Oberon as a 'dead world', especially in view of its crater statistics.

Figure 5.24 is a relative size–frequency distribution plot for craters on the five major moons of Uranus. This shows that both Oberon and Umbriel have dense populations of large impact craters comparable with the lunar highlands and heavily cratered terrains elsewhere in the solar system. By analogy with the situation at Saturn, this distribution is attributed to an early, population I, bombardment. In contrast Titania and Ariel show a population II style of cratering, whereas Miranda is intensely cratered with a peculiar distribution of its own. It is generally believed that the population I bombardment recorded on Oberon represents the equivalent of late heavy bombardment by post-accretional debris. Calculations show that gravitational focusing of this impacting population would have resulted in a greatly enhanced rate of cratering on the inner satellites, to the extent that Umbriel and Ariel are likely to have been struck at least once during this period (say 4.0 to 3.5 billion years ago) by a projectile with sufficient energy to have completely disrupted them. The same statistical argument suggests that Miranda would have been disrupted and re-accreted five times, and when Miranda is discussed (in the next chapter) it will be seen that it does apparently bear the scars of such an event.

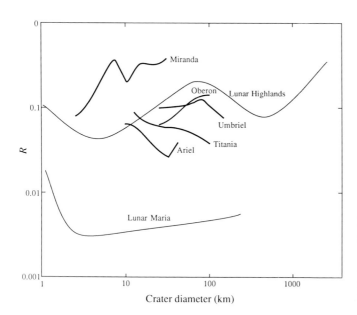

Fig. 5.24. Relative size–frequency distribution plot showing the crater populations of the most densely cratered regions on the satellites of Uranus.

5.8 Umbriel

Umbriel is smaller than Oberon, and, in terms of size and mass, is the twin of Ariel, although their geological histories have been very different. Umbriel's population I crater record has already been remarked upon. This implies that if Umbriel was disrupted by a major impact during the population I bombardment, as statistics extrapolated from Oberon suggest is likely, then it must have happened early enough for Umbriel to have re-accreted in time for its new surface to experience a significant duration of continuing population I bombardment.

As its name implies, Umbriel is a dark world, with an albedo of slightly less than 0.2. The morphology of the craters on its surface shows little or no evidence of any viscous relaxation (Figs 5.25 and 5.26). Unlike Oberon, or indeed the other satellites of Uranus, Umbriel has no craters with bright rays or other pale ejecta blankets. A possible exception is a bright annular feature seen near the limb in Fig. 5.25. The overall albedo is consistent with the radiation-darkening of methane with age, although the uniformity of the albedo is surprising. Impacts would be expected to have excavated cleaner, paler material from depth, or, if the outer layer is completely uniform to below the depth excavated, at least to have produced a rather paler ejecta simply by fragmentation and pulverization. These problems have lead to speculation that Umbriel has recently been given a surface dusting by dark material of unknown provenance that has hidden whatever surface markings there were previously. Suggestions as to the origin of the dark material include a global mantle of dark dust from a recent impact (a 10 km crater would be sufficient), or an explosive volcanic eruption, although this is not compatible with any observed aspect of Umbriel's surface morphology. Alternatively, perhaps the radiation-darkening of methane is entirely

Fig. 5.25. *Voyager-2* image of Umbriel, showing its heavily cratered surface. The large crater with a prominent central peak near the upper left is 110 km in diameter. Umbriel is darker than Oberon (Fig. 5.22), but this is not apparent in comparing these images because of subsequent enhancements.

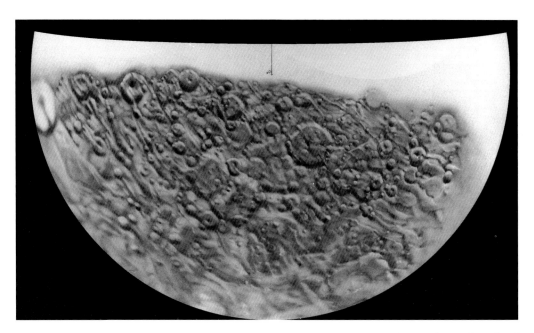

Fig. 5.26. Shaded relief map of the southern half of the Uranus-facing hemisphere of Umbriel, covering much the same area as the image in Fig. 5.25.

responsible for the uniformity of Umbriel's surface, and the persistence of albedo features on the other satellites of Uranus demonstrates the absence of methane in certain regions or at certain depths on those worlds, but its ubiquity on Umbriel.

With Umbriel we have come to the last of the 'dead worlds'. In the next chapter we will examine a comparable number of worlds, many of them similar in size, mass and bulk composition to those in this chapter, but where the record of early cratering has been partly overprinted by the indisputable traces of tectonism and volcanism.

6

Recently active worlds

The fashion of this world passeth away
I Corinthians 7:31

With this class of worlds we come to those where a geologist can feel truly at home. Indeed, they excite the curiosity of just about anyone who has seen the images, because no longer do we have to be content with counting and measuring craters and pondering the significance of a few dubious fractures. These are the worlds where there is abundant evidence of tectonics on global and local scales, and of vast volcanic outpourings. However, even the youngest terrains formed in this way bear the scars of subsequent impact cratering; this means that they post-date the late heavy bombardment (or its outer solar system equivalent), but are none the less very ancient. If they occurred on the Earth these later terrains would be called neither young nor recently active (indeed most of them probably pre-date the oldest surviving terrestrial terrains), but on worlds such as these, without air and running water or younger episodes of tectonics and volcanism, most of their original characteristics have been preserved. We begin with Ganymede (Jupiter), which has been the most intensively studied example of its class. The other satellites included in this chapter are Dione and Tethys (Saturn), and Ariel, Titania, and Miranda (Uranus).

6.1 Ganymede

Ganymede is the largest and most massive of all the planetary satellites. It is bigger than Mercury, although less than half its mass, and is more than twice the diameter and ten times the mass of Pluto (Table 1.1). Were it not in orbit around Jupiter, there is no doubt that Ganymede would be regarded as a respectable planet in its own right. Actually it is not very much bigger than Callisto, and the question of why Ganymede is so clearly a differentiated body with abundant evidence of tectonics whereas Callisto shows no tectonism and may not be differentiated at all is one of the major unresolved issues in planetary science.

The wispy markings that can be seen on Ganymede from Earth using large telescopes and that were imaged by *Pioneer-10* (Fig. 2.7) contain the key to Ganymede's strange geology. The higher resolution images obtained during the two *Voyager* encounters show that the globe is divided into roughly equal proportions between a dark terrain that occurs as polygonal regions and is heavily cratered, and a less heavily cratered bright terrain in the form of bands that separate the dark terrain polygons and in places penetrate the dark terrain in the form of wedges (Fig. 6.1). The bright terrain is clearly younger than the dark terrain. It is usually strongly grooved, and provides unequivocal evidence of one or more episodes of widespread-to-global tectonism and volcanism. However, even the bright terrain is fairly heavily cratered, and both it and the dark terrain are peppered with younger craters with bright rims and ejecta blankets. With so many prominent features to name, the stock of available names from the Ganymede myth was quickly depleted, and the majority of names on Ganymede have been drawn from the mythology of ancient Near Eastern civilizations.

6.1.1 *The dark terrain*

We shall briefly consider the dark terrain first, as this has many similarities with, and some notable differences from, the surface of Callisto. In general, Ganymede shows the same sorts of craters as Callisto, except that their topography has on the whole become even more subdued by viscous relaxation. The size–frequency distribution of craters on the dark terrain is more or less the same as on Callisto (Fig. 6.2). However, whereas craters up to about 100 km in diameter are common on the dark terrain, larger craters are rare. In their place are ancient pale circular patches of subdued relief 100–300 km in diameter (Fig. 6.3) that resemble the bright palimpsest in the centre of the Valhalla basin on Callisto and are referred to by the same term. Each palimpsest is surrounded by concentrations of smaller craters that are almost certainly secondary craters produced by blocks of ejecta from the main crater-forming incident. Ganymede's palimpsests may be the traces of viscously relaxed major impact structures, as inferred on Callisto. An alternative explanation is that the impacts that produced them broke right through the lithosphere (which would have to have been thinner than about 10 km at the time), and that the palimpsests represent extrusions of warm ice or a slushy ice–water mixture, which was able to escape buoyantly from the asthenosphere and then spread symmetrically across the surface.

Figure 6.2 shows that the crater distribution curve for the dark terrain on Ganymede has much the same shape as the curve for Callisto. This may indicate that, as expected, these two worlds were bombarded by the same population of impactors. An important difference, however, is that Ganymede's curve lies below that for Callisto, showing that the overall crater density on Ganymede's dark terrain is less than the crater density on Callisto by a factor of about three. Other things being equal, Ganymede ought to have more craters than Callisto as a result of the gravitational focusing of incoming projectiles by Jupiter's gravity. The best way to account for the relatively low number of craters on Ganymede's most ancient terrain is if it simply has a younger crater-retention age. This could be a result of more effective viscous relaxation on Ganymede or of a global resurfacing (perhaps melting) event. Either of these could be explained by the extra heat generated within Ganymede due to greater radioactive, tidal, or accretional heating, as a consequence of Ganymede's being slightly denser and larger than Callisto. Studies of the relaxation of crater topography suggest that Ganymede's lithosphere thickened from about 10 km at the time of the earliest craters preserved (4.0 billion years ago?) to about 35 km by the time the first bright terrain was formed (3.8 billion years ago?).

There are no well-preserved multiringed basins on Ganymede, but there are at least

Fig. 6.1. Shaded relief airbrush map of Ganymede's leading hemisphere (*this page*) and trailing hemisphere (*opposite page*), based on *Voyager* images. Changes in the degree of detail shown reflect the variations in resolution coverage (see Fig. 3.6). These views show clearly the division into dark and bright terrains, and the yet brighter ejecta from the younger craters.

four systems of roughly concentric arcuate furrows, consisting of troughs approximately 10 km wide with raised rims spaced about 50 km apart. These can be traced for distances of several hundred kilometres, and part of such a system is well seen in the lower half of Fig. 6.3. It has been proposed that these systems are of tectonic origin and record an episode of global expansion, but their concentric nature and analogy with Callisto strongly suggests that they are due to impacts. The fact that most large craters overprint any furrow

that intersects them shows that the furrow systems were originated relatively early in the history of the dark terrain, in which case any central impact structure would be expected to have relaxed to the palimpsest stage or beyond. The most complete furrow system does appear to have a faint palimpsest at its centre. Unfortunately, however, we have no knowledge of the centres of the other three examples, because of a combination of resurfacing by the bright terrain and gaps in the high resolution coverage by the *Voyager* imaging systems.

Whatever caused them to form, the present morphology of individual furrows can be

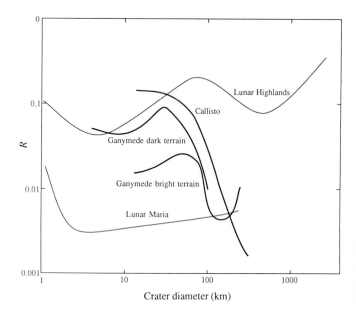

Fig. 6.2. Relative size–frequency plot comparing the crater populations on Ganymede and Callisto.

fairly simply explained, as shown in Fig. 6.4, by analogy with the topographic adjustments experienced by terrestrial fault-bounded valleys: a trough forms as a graben, by down-dropping of the central block; viscous relaxation in response to isostatic forces then causes the rims and the centre of the basin floor to bow upwards and the edges of the floor to be depressed. On Ganymede, this modified topography is likely to persist indefinitely, or at least until obliterated by the superposition of craters, whereas on Earth the raised rims of analogous structures experience rapid erosion, leading to continued isostatic uplift and further erosion of the rims, so the process is liable to continue until the graben has been filled with sediment and the topographic anomaly has been erased.

The observation that the dark terrain is broken by bands of bright terrain raises the question of whether the blocks or plates of Ganymede's lithosphere separated by these fractures were ever decoupled from the interior and thus able to slide around in a manner analogous to plate tectonics on the Earth. Many geologists would dearly love to be able to study the effects of plate tectonics on another world, but unfortunately Ganymede is not the place. The ancient furrow patterns and craters within the dark terrain that are truncated by the bright terrain show no major offset from one block of dark terrain to the next. There is also a slight (and contentious) asymmetry in the crater size–frequency distribution of the dark terrain between the leading and trailing hemispheres. This indicates that, as can be argued for Callisto, there have been more impacts in the leading hemisphere, and the preservation of such a trend demonstrates that the lithosphere must have been immobile with respect to Ganymede as a whole throughout the time recorded by the cratering history. Thus Ganymede's lithosphere seems to have been fixed in place since before the start of the geological record. This rules out plate tectonics and any other

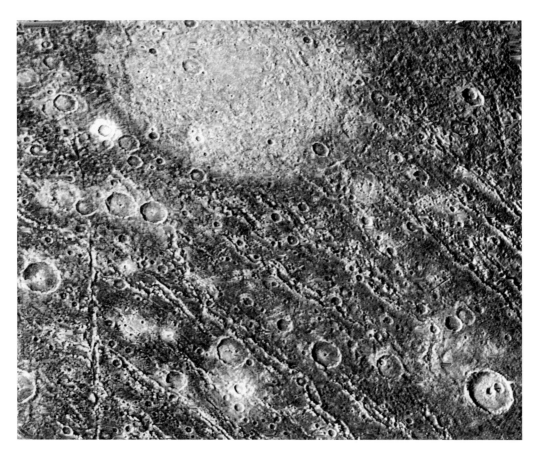

Fig. 6.3. *Voyager-2* image covering a dark terrain region on Ganymede, about 700 km across. The 350 km diameter bright palimpsest near the top edge is Memphis Facula. This is either the relic of a large impact crater that formed when the lithosphere was sufficiently warm and thin to allow it to become flattened under its own weight, or a region covered by the extrusion of warm ice when an impact penetrated the still-thin lithosphere. Virtually all the craters in this image are topographically relaxed. An ancient arcuate furrow system trends obliquely across the image; most of the craters and the palimpsest overprint these furrows, and so the furrows are clearly very old.

(a)   (b)

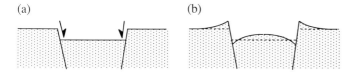

Fig. 6.4. Cross-sections to show topographic relaxation of a graben, which may explain the raised rims on the furrows on Ganymede. (a) Immediately after formation and (b) after viscous relaxation of topography.

process of regional motion or rotation of terrain blocks. To determine the kind of processes that did go on, we need to examine the bright terrain.

6.1.2 *The bright terrain*

Ganymede's bright terrain is where the geology begins in earnest. It is clearly younger than the dark terrain, as can be seen by truncation relationships wherever they abut (Fig. 6.5), and by the fact that the bright terrain has a lower crater density (Fig. 6.2). In fact, the crater density on the bright terrain varies from place to place (even after allowing for the possible asymmetry due to Ganymede's orbital motion), which suggests that the bright terrain was emplaced over a period of the order of half a billion years, probably at some time before 3 billion years ago.

Most of the bright terrain is occupied by belts of sub-parallel grooves (Fig. 6.6), and indeed the bright terrain is alternatively known as the grooved terrain. In some places the bright terrain is smooth, notably where it overlies the dark terrain with an indistinct boundary, as in places on Fig. 6.5. This suggests that the smooth unit was emplaced as a low viscosity fluid, probably water or a mush of ice crystals lubricated by a small proportion of melt. In most places, however, the boundary between the bright and dark terrains is sharp, and where the bright terrain occurs as a belt cutting between areas of dark terrain, the grooves in the bright terrain run parallel to the boundary. Some sets of grooves can be seen to truncate older sets, and there are extensive regions of bright terrain

Fig. 6.5. An isolated, elevated remnant of Ganymede's dark terrain, about 120 km across, surrounded by bright terrain. Dark terrain also occurs in the background. Topographic features in the dark terrain are truncated by the mostly smooth texture of the bright terrain in this area.

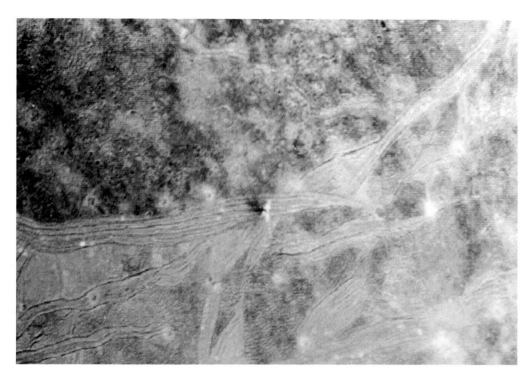

Fig. 6.6. Bright grooved terrain on Ganymede, occupying most of the lower portion of the image. Several generations of grooves are demonstrated by their cross-cutting relationships. Dark terrain occupies the upper part of the image, and also occurs as fragments within the grooved terrain. View about 2000 km across.

consisting of several generations of grooves, some of which meet the boundary with dark terrain at steep angles. Occasionally, one set of grooves overprints another, resulting in a reticulate, criss-cross pattern.

The age range of the bright terrain as determined by crater statistics is compatible with the complex emplacement history of segments of grooved terrain. The youngest grooved terrain appears to consist of single grooves and pairs of grooves, and there are also some groove-like features within the dark terrain that have no distinct albedo contrast (Fig. 6.7). The bright terrain then appears to be the result of a prolonged phase of global activity that built up to a climax and then waned, with single grooves representing the earliest and latest activity. The most probable origin for the bright terrain, especially the belts of grooves, is that it is a result of global expansion produced by internal differentiation. Had Ganymede accreted homogeneously, ice phase changes could result in up to a 7 per cent increase in surface area if all the rocky fraction were able to segregate into the centre. This is an extreme model, and as the bright terrain occupies something like half of Ganymede's surface, it obviously cannot simply represent extra surface area generated by expansion. Instead, it must be fresh material extruded on to the surface in response to extensional

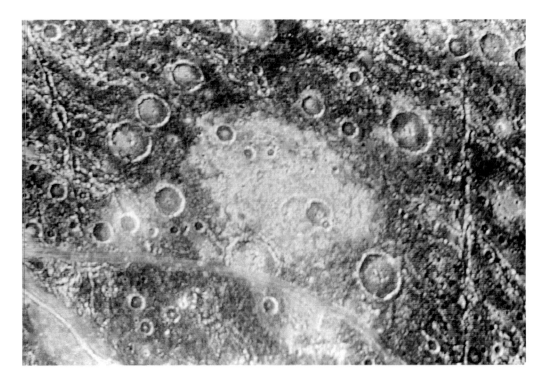

Fig. 6.7. A 300 km wide extract of a *Voyager-2* image, partly overlapping the region shown in Fig. 6.3. Just to the south of the palimpsest near the centre, the southern rim of a 60 km diameter crater has been destroyed by the formation of a narrow, double-grooved, belt of bright terrain (one of the youngest bright terrain features on Ganymede). Two groove-like features run north–south in the dark terrain (possibly representing an early stage of the activity that lead up to bright terrain formation), and it can be seen where the westernmost of these is cut by the bright belt that there is no sideways offset.

cracking of the lithosphere. This process would be encouraged by thermal contraction of the lithosphere as it cooled, at the same time as phase changes and/or internal differentiation caused the volume of the globe to increase. A phase change that could be responsible is the conversion of ice V to ice II as the internal thermal gradient decreased (Fig. 6.8).

If the bright material was extruded as water, brine, or an icy crystal mush, it would be encouraged to rise through fractures and spill out on to the surface provided that the ice in the lithosphere was made slightly denser than the rising material by the incorporation of rocky fragments (as seems likely). The material extruded from below would be more reflective than the dark terrain because it would contain fewer rock fragments and less dust. There are three different general models that might account for the formation of belts of bright terrain. These, and a variant of the third model, are illustrated in Fig. 6.9.

In model (a) the belt of bright terrain occupies a gap produced by the tearing apart of blocks of lithosphere on either side; the emplacement of the bright material would be

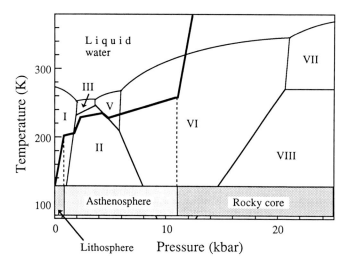

Fig. 6.8. Ice phase diagram for a fully differentiated Ganymede. The heavy line shows the thermal gradient at the time of bright terrain formation. The lithosphere consists of ice I with some rocky contaminants, whereas the covecting asthenosphere contains ices I, II, V, and VI. Parts of the asthenosphere in the ice II region are only about 20 K below the melting point for pure ice, and could begin to melt in the presence of dissolved salts or volatiles, thus acting as a source of the fluid that was extruded to form the bright terrain. As the interior cools, ice V will be converted to ice II, resulting in a slight net expansion. At the present day, the lithosphere is much thicker, and extends well into the ice II stability field.

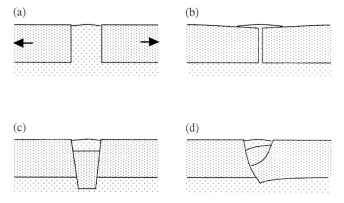

Fig. 6.9. Alternative models to account for the formation of Ganymede's belts of bright terrain by the extrusion of water or relatively pure ice slurry from the asthenosphere, in response to extension. See text for discussion.

analogous to the formation of an ocean basin on Earth by sea-floor spreading. There are two main objections to this model that seem to rule it out completely. One is that the implied extension is far greater than can be accounted for, in the absence of any evidence of a nearly equivalent amount of lithospheric compression elsewhere on Ganymede. The second is that although pre-existing craters may be truncated by the edge of a belt of bright terrain (e.g. Fig. 6.7), there is only one good example of a crater seen to be split apart by such a belt; in other words there is no place where the dark terrain on either side of a bright belt can be fitted back together again, in a manner analogous to the way in which Africa and South America fit neatly if the South Atlantic Ocean is closed.

In model (b) the bright belt is produced by flooding above a tensional crack. There are sound objections to this model as well; the straightness of the edges of the bright belts and the lack of flooded craters at the edges of these belts point strongly in favour of structural confinement by faults, whereas simple flooding would tend to result in more tortuous boundaries. In addition, the bright material in the grooved belts is evidently at least 10 km thick, because craters many tens of kilometres in diameter lie within this terrain without excavating dark material from beneath (Fig. 6.10).

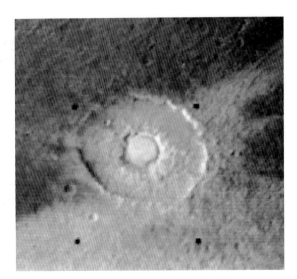

Fig. 6.10. Detail of a topographically relaxed 100 km diameter central pit crater stradling the boundary between Ganymede's bright and dark terrain. It can be seen that the inner wall and rim deposits of the crater are dark where the crater was excavated in dark terrain and paler where the crater cuts into bright terrain, which indicates that each of these units must extend to a depth of at least 10 km.

The model that best fits the requirement of structural confinement (to account for the straight edges of the belts), thickness of the bright material within the belts, and the small amount of extension possible is that the edges of a bright grooved belt are defined by faults, between which the lithosphere has dropped. The resulting trough could then be flooded by upward-escaping fluids. Model (c) shows this trough as a classic graben, bounded by two faults of equal importance, each of which reaches the base of the lithosphere. The variant shown in (d) is more in line with discoveries on Earth during the late 1980s that revealed that terrestrial grabens are more commonly formed by extension across a single dominant fault and that the opposite side of such a rift valley is delimited by a less significant fault, that joins the major fault at depth.

The nature of the flooding process in the grooved terrain would undoubtedly cast light on the problem of how the bright belts formed, if only we could understand it. Unfortunately the topographic profiles across the grooved terrain are not well constrained, and in any case they have probably been modified by local tectonic tilting and viscous relaxation subsequent to flooding. The morphology of the grooves cannot therefore be used to tell us anything about the physical properties of the flow material that was extruded; the grooves and the intervening ridges could be original tectono-volcanic constructs but they could equally well be the result of subsequent subsidence or rotation of underlying blocks. A few of the possible mechanisms for groove formation are shown in Fig. 6.11.

(a)

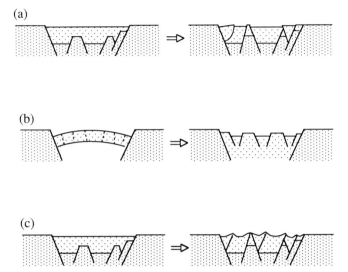

(b)

(c)

Fig. 6.11. Some of the ways to account for the formation of ridges and grooves within belts of bright terrain on Ganymede, each beginning with the flooding of a fault-bounded valley. (a) Ridges and grooves formed by differential subsidence and rotation of blocks; (b) upward bending of the surface caused by expansion as the ice freezes gives rise to extensional cracks, which, after subsidence, produce a ridge and groove topography; and (c) late-stage extrusion of viscous melt along fault-controlled fissures to construct ridges volcanically.

As no one has yet come up with a wholly convincing explanation of the morphology of Ganymede's grooved terrain in terms of an extensional origin, it is tempting to look for alternative mechanisms. Some grooves on Ganymede can be traced for about 1000 km, which is considerably greater than the length of any continuous volcanic fissure or extensional fault on Earth. However, faults across which the relative movement is sideways do extend for this sort of distance on Earth. The San Andreas fault that is responsible for many of the earthquakes in California, USA, is a strike-slip fault of this type. In spite of this, the fact that the remnants of ancient furrow systems in regions of dark terrain that are separated by grooved terrain remain roughly concentric argues strongly against major strike-slip motions, as does the leading–trailing asymmetry in dark terrain cratering. The most that can be said is that with all the extension that went on, there must inevitably have been *some* sideways motion as well, if only to satisfy the geometric constraints imposed by the disruption of the lithosphere into several blocks. The *Voyager* images show a few debatable examples of sideways displacements on Ganymede of up to a few tens of kilometres, but greater displacements than this can almost certainly be ruled out.

6.1.3 *Subsequent history*

Whatever the mechanism by which the grooved areas and the rest of the bright terrain developed, the morphology of superposed craters shows that the first ones to form on this terrain became extremely flattened due to viscous relaxation, although later craters have much the same shape as craters of similar age on the dark terrain. This tells a story of cooling and stiffening of the lithosphere in the bright terrain as it aged.

The thickness of Ganymede's lithosphere today is probably about 300 km and there is no evidence of changes in global volume since the formation of the bright terrain. This is taken to show that the potential expansion as ice V changed to ice II with cooling was offset by the potential contraction as ice I changed to ice II, in conjunction with general thermal contraction.

The most noteworthy event on Ganymede in post-bright terrain times has been the formation of an impact basin 275 km in diameter that has been named Gilgamesh (Fig. 6.12). This is evidently younger than any of the multiringed structures on Callisto, clearly having formed when the lithosphere had already reached a considerable thickness. This makes it a closer analogue to the major impact basins on the Moon (e.g. Fig. 5.12), Mars and Mercury, and the blocky mountainous region surrounding the structure is the most rugged terrain known on any of the outer three galilean satellites.

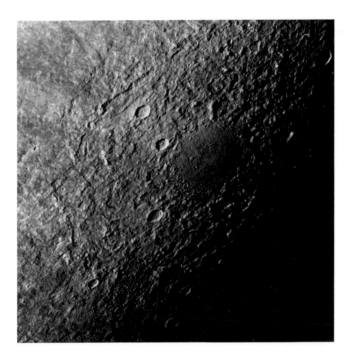

Fig. 6.12. A 650 km wide *Voyager-2* image showing Gilgamesh, a young, morphologically fresh giant impact structure on Ganymede. The central smooth area, towards the upper right of this view, is about 150 km in diameter. The most conspicuous concentric ring, which is 250–300 km from the centre of the basin, is marked by an irregular inward-facing scarp up to 1.5 km high. Most of the craters in the lower left are secondary, having been produced by ejecta thrown out by the main impact.

Although Ganymede is the foremost example of a recently active world, as we shall see shortly it is by no means typical. Nowhere in the solar system is there anything that bears close comparison with Ganymede's grooved terrain. Most of the other recently active worlds are characterized by having two or more cratered terrain units of different ages transected by tectonic features, mainly in the form of broad fault-bounded valleys. We shall continue our survey with Dione and Tethys, two of the moons of Saturn.

6.2 Dione

Among Saturn's moons, Dione (Fig. 6.13) exhibits the most abundant signs of internal activity, with the exception of Enceladus, which is described in the next chapter. Dione is near the middle of the size range, but has the highest well-determined density of Saturn's satellites apart from Titan (Table 1.1). Like Rhea it has a bright icy surface (albedo 0.5), and its leading hemisphere is distinctly brighter than its trailing hemisphere; the opposite situation to that found at Iapetus. The best images we have are at a resolution of no better than 2 km, but the global coverage is rather more complete than for Rhea.

6.2.1 *Terrain types and tectonic features*

Three major terrain types have been identified on Dione (Fig. 6.14); cratered terrain (ct) with relatively numerous craters in excess of 20 km diameter, cratered plains (cp) with an intermediate crater density, and smooth plains (sp) with a low crater density. The boundaries between these terrain units are indistinct and are not marked by any major surface structures, but they can usually be defined to within a crater-width or so. The cratered terrain is significantly poorer in craters 20 km in diameter than even the least densely cratered regions on Rhea, and the variation in crater density between Dione's terrain units is much greater than the differences between regions on Rhea. It appears that the heavily cratered terrain must have been resurfaced towards the end of population I bombardment and that the two younger terrain units represent resurfacing episodes that occurred during population II times.

   Several kinds of tectonic feature can be mapped on Dione. There are pale wispy streaks in the poorly imaged regions (Fig. 6.15), some of which can be traced into areas imaged at higher resolution where they are shown to be troughs about 8 km in width, such as Palatine Chasma (Fig. 6.16a). These troughs are of a variety of ages, as can be seen by their variable overprinting relationships with craters. Some rather degraded troughs can be made out in the cratered terrain and cratered plains, and better preserved troughs are visible in parts of all three terrain units. The troughs are usually regarded as grabens, and in places where two or more troughs run closely parallel to one another there appears to be a classic horst and graben situation, with a linear up-faulted block between two down-faulted troughs (Fig. 6.16b).

   Troughs similar to Palatine Chasma are common further north and there is also one of a rather different form, being 20–30 km wide and having irregular, scalloped walls (Fig. 6.14). Some of the more sinuous narrow troughs, such as those in the smooth plains in the north of Fig. 6.13, can be seen to split and terminate in pit craters. There are also several chains of coalescing pit craters, which are best seen in the plains areas. These may be collapse features, in which case they can be regarded as transitional forms of the troughs. Alternatively, it has been suggested that the troughs on Dione were the sites of volcanic eruptions related to the plains-forming events; if so the pit craters are likely to have been explosive vents.

Fig. 6.13. *Voyager-1* mozaic covering the Saturn-facing part of the leading hemisphere of Dione. The most prominent trough in the upper left is Latium Chasma, and the largest crater in this view, Aeneas, has a diameter of about 200 km. Palatine Chasma (see Fig. 6.16) is visible near the southernmost limb.

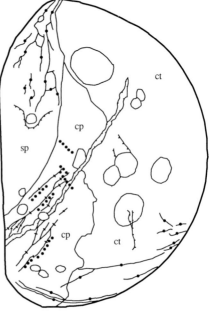

Fig. 6.14. Terrain units and major landforms on Dione. The map covers the same area as the image in Fig. 6.13. In order of decreasing crater density the terrain units are: *ct*, cratered terrain; *cp*, cratered plains; and *sp*, smooth plains.

- Scarp
- Irregular-wall trough
- Trough
- Ridge
- Crater chain

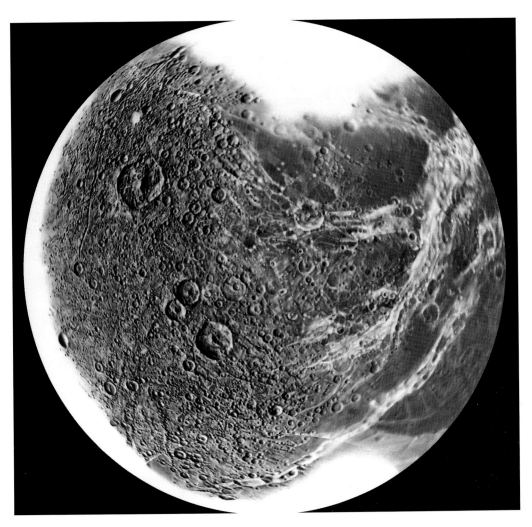

Fig. 6.15. Shaded relief map of Dione's Saturn-facing hemisphere. Some of the pale wispy streaks in the poorly imaged trailing hemisphere (*right*) can be traced into troughs in the leading hemisphere (*left*).

Dione also has a series of broad, convex ridges up to half a kilometre high within the cratered plains unit (Fig. 6.14) that are parallel to some of the crater chains and the irregular walled trough, suggesting a common structural control for all three types of feature. The ridge morphology could be volcanic or due to compression. Evidence of compression in Dione's history is also provided by a few scarps up to 100 km long, notably in the heavily cratered terrain, that are mostly linear and less than 1 km high.

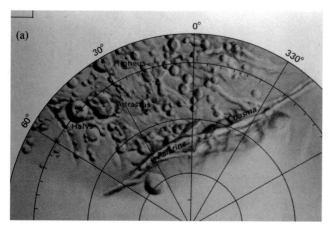

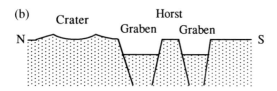

Fig. 6.16. (a) Part of a shaded relief map of Dione's south polar region showing Palatine Chasma, a relatively fresh-looking trough that can be traced for over 600 km (see Fig. 6.15). Vestiges of older troughs can be seen particularly well between 330° and 0°. (b) A cross-section through a double portion of Palatine Chasma along the 30° meridian showing a horst and graben interpretation of its structure. In detail, each trough appears to be U-shaped in cross-sectional profile, probably as a result of mass-wasting of the slopes.

6.2.2 *Geological history*

It has to be admitted that the evidence we have for describing Dione's geological history is far from adequate; there are no clear signs of volcanism, each of the tectonic features described above could be attributed to several origins, and the relative ages of different features are not well-constrained. However, the images from Dione can be interpreted in a way that is broadly compatible with its supposed composition and its modelled thermal history, as outlined below.

The radiogenic heat supply within Dione would probably result in a somewhat similar thermal history to that shown for Rhea in Figure 5.17, with Dione's smaller mass being partly offset by the greater ratio of rock to ice implied by its density. Equivalent models for these two satellites show that convection would be at its most intense in Dione about 180 million years after accretion and in Rhea about 140 million years after accretion, but that convection would stop sooner in Dione, at about 2.7 billion years after accretion. If resurfacing occurred during the most intensive stage of convection, the later peaking of Dione's convection may be adequate to explain why its surface mostly post-dates the population I bombardment, whereas resurfacing on Rhea went on before the population I bombardment had declined, resulting in a more heavily cratered surface with poorly preserved signs of tectonics (at least in the well-imaged areas). The declining stage of convection would have gone on under an increasingly thick lithosphere, so that the more prolonged convection on Rhea would not have been able to drive any resurfacing process.

The two plains units on Dione may represent regions of heavily cratered terrain that were flooded by eruptions of $NH_3 \cdot 2H_2O$ melt (see section 4.2.3) at different times. If so, it was presumably a low viscosity melt because it spread over a large area (probably in many stages) and left no traces of any steep flow margins. Alternatively, the eruptions could have been explosive, in which the sudden vaporization of methane or ammonia threw out clouds of icy particles, which would have settled to the surface as a form of volcanogenic snow. A problem with the volcanic theories is that there are no traces of flooded or buried craters showing through from the underlying surface, and an alternative model is that those regions now occupied by plains units were subjected to higher heat flow, at which time all pre-existing craters large enough to be seen on the images (about 2 km and above) were removed by viscous relaxation. Thus the plains units may not represent volcanically flooded areas at all, and the various terrain ages are then due to the maintenance of anomalously warm tracts of lithosphere above regions of convective upwelling.

The timing of either the volcanic or thermal resurfacing processes would have been similar, occurring early on in Dione's history when convection and partial melting were at their most vigorous and reached nearest to the surface. The extrusion of large volumes of melt would also be favoured by global expansion, which would put the lithosphere under tension and thus open up pathways by which the melt could reach the surface. The results of one attempt to predict the size of Dione over time are shown in Fig. 6.17; unlike Rhea, Dione is too small for ice II to exist even at its centre, so the only possible causes of volume change are thermal expansion or contraction, melting, freezing, and differentiation. On the model shown, the initial episode of increasing radius is due to thermal expansion of the ice as radiogenic heat built up inside. This expansion left its mark in the structure of the lithosphere, being responsible for initiating all or most of the troughs, although many of them may have been reactivated later.

An analysis of the orientation of the linear features in both the well and poorly imaged areas shows that they tend to run either to the north-east or to the north-west except in the

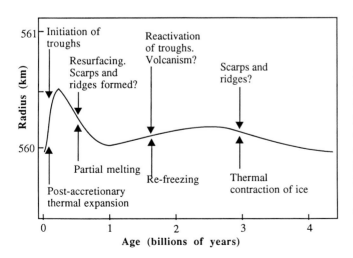

Fig. 6.17. Model for changes in Dione's radius over time. The timing and amount of the initial phase of expansion and contraction are strongly dependent on the amount and relative density of the partial melt. The causes of the size changes are noted below the curve, and possible volcanic and tectonic consequences are shown above it.

polar regions where they tend to run east–west. This is precisely the pattern predicted for fractures in a lithosphere where the directions of stress are controlled by tidal despinning of the satellite's rotation. It seems reasonable that Dione's linear features developed while it was being brought into captured rotation, and that the thermally predicted global expansion caused the lithosphere to fracture in directions that were dictated by the tidal despinning stress pattern.

According to the model, the result of this heating was a large amount of partial melting, and as the liquid would have been denser than ice I, this lead to global contraction. This could have caused some of the scarps and ridges, which seem to be compressional features. The presence of a large volume of melt would have enabled volcanic resurfacing to take place (forming the plains units) by the passage of melt up the stress-generated fractures, although the compressive nature of the stress in the lithosphere at this time, a result of the global contraction, would have hindered this. Alternatively, the strong convection facilitated by the high temperatures and the presence of the melt could have caused resurfacing by viscous relaxation. At this time, the rocky component would have had the opportunity to segregate downwards to form a core perhaps 200 km in radius.

The next phase in Dione's history was when melt generation had ceased and it began to re-freeze. This resulted in a prolonged episode of global expansion because of the density contrast between melt and solid, and it is possible that any volcanism that did occur went on during the early part of this phase, when melt still existed in considerable quantities and the lithosphere was once again under stress. Some troughs, especialy a few that have raised rims, and chains of coalescing craters could have acted as volcanic fissures, feeding either lava or explosive eruptions.

When the last of the partial melt in the asthenosphere had re-frozen, the final phase of Dione's thermal history would have begun, in the form of slow global contraction caused by thermal contraction of the ice. It is unlikely that this phase has any major surface manifestations, although some of the compressional features could date from this time.

Note that the model just presented assumes that Dione as we see it has not been reaccreted after destruction by a major impact. Such an event could explain the general lack of population I craters, but the implications of a young age are severe in greatly reducing the extent of radiogenic heating that would have been possible. That issue aside, it is almost certain that the above interpretation of Dione's history is flawed, at least in many of its details. Images with higher resolution would undoubtedly help, but we may not get significantly closer to the truth until expeditions have landed on Dione and have been able to examine the composition and structure of the surface material at first hand. Each of the moons of Saturn displays its own variation on the theme of thermal-related volume changes, tectonics, and resurfacing. Another permutation is seen in the next section where Tethys, the twin of Dione, is discussed.

6.3 Tethys

Tethys is almost identical in size to Dione, although its lower density shows that it has a smaller proportion of rocky material, and thus is likely to have experienced less radiogenic heating. Indeed, Tethys does exhibit fewer signs of internal activity and has elements reminiscent of Dione, Rhea, and Mimas mixed together on its surface (Figs 6.18 to 6.20).

All parts of Tethys are densely cratered, much of it showing clear signs of dense population I cratering with most of the larger craters being highly degraded by viscous relaxation. Part of the trailing hemisphere is covered by a lightly cratered plains unit like that on Dione, which is a reasonable sign of local resurfacing after population I time. It seems that this resurfacing occurred by volcanic flooding, because some old, larger craters have survived in this terrain, but have low rims suggesting that they were almost, but not quite, overwhelmed by the volcanic outpourings, whereas all the pre-existing smaller craters were buried by the flood (Fig. 6.18). The edge of the cratered plains unit has an intriguing appearance, but unfortunately all but one of the scheduled highest resolution (*ca.* 2 km) images of Tethys were lost (as a result of the jamming of one of the axes of *Voyager-2*'s scan platform described in Chapter 3), so details such as this remain uncertain.

The surface of Tethys is marked by two giant features that make it stand out amongst Saturn's family of moons. The first of these is the crater Odysseus (Fig. 6.19), named after the central character of Homer's *Odyssey*, which at a diameter of about 440 km (about 40 per cent of the diameter of the satellite) is the largest crater in the Saturnian system. Unlike the crater Herschel on Mimas, which is of only slightly smaller magnitude in relation to the world on which it occurs, the topography of Odysseus has relaxed to a large extent. Its floor has rebounded by several kilometres, and the one *Voyager* image with Odysseus on the limb shows that the floor is convex with roughly the same radius of curvature as the satellite as a whole. The crater walls are low and the central complex is no more than a shadow of the impressive peak it must have been shortly after formation.

The second giant feature is an enormous trough, Ithaca Chasma, that extends at least three quarters of the way round the globe (Fig. 6.20). It has a width of up to 100 km and reaches about 3 km in depth. The inner walls show tantalizing evidence of terracing and other structures, but again we are thwarted by the poor resolution of the images. Poignantly, for Ithaca was Odysseus's home that he suffered such long hardship trying to return to, Ithaca Chasma falls along a great circle with the crater Odysseus lying at almost the maximum possible distance away, near one of the poles to this great circle. This has prompted the suggestion that the Odysseus impact was responsible for the formation of the chasma. The most probable mechanism is that the chasma opened up as a direct response to the viscous relaxation of the giant crater. To open the chasma in this way, the viscous flow must have been inwards towards the crater *everywhere*, instead of inwards in the near region but outwards in regions further away, as for major impacts whose size is less in relation to the diameter of the globe (see Fig. 5.13).

In view of the size ratio between the crater Odysseus and Tethys itself, the viscous relaxation of Odysseus probes the internal structure of Tethys very deeply, and it can be shown that for the viscous flow to have been unidirectional, as suggested by the formation of the chasma, any silicate core that Tethys may possess could have occupied no more than about 20 per cent of its radius at the time the relaxation occurred. This compares with a core radius of about 47 per cent if Tethys were fully differentiated. In addition, the

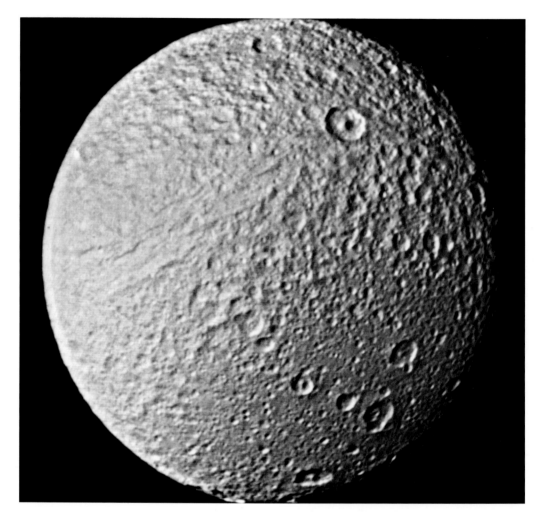

Fig. 6.18. The best global view of Tethys, from *Voyager-2*, with a resolution of about 4 km. Most of the area covered is heavily cratered terrain, but the lower right of the image is occupied by a considerably less densely cratered unit (cratered plains). There are intriguing hints in the image that this terrain was emplaced by flooding. Ithaca Chasma (see Fig. 6.20) runs diagonally downwards to the left from the prominent crater Telemachus (*near the top right*), but is not well seen because of the high sun angle.

unidirectional flow puts an upper limit on the internal thermal gradient of about 0.01 K km^{-1}. These are almost the only constraints we can put on the internal structure and differentiation of any of Saturn's satellites using direct observational evidence, and even these remain open to dispute. Even if we accept the genetic link between Ithaca Chasma and Odysseus, the lack of apparent relaxation of the chasma's topography

Fig. 6.19. An enhanced image showing the highly relaxed crater Odysseus (*lower right*), dominating the surface of Tethys.

suggests that it may have been reactivated at a later stage. Furthermore, the crater density on the floor of the chasma is slightly less that of the cratered plains, showing that the present floor, if not the chasma as a whole, is the youngest surface yet identified on Tethys. Even so it is old, probably not significantly younger than the cratered plains units that occur on both Tethys and Dione.

We now move out to Uranus, where there are other worlds whose surfaces are marked by major troughs, with the added interest that the link between tectonism and volcanism is much clearer.

6.4 Ariel

As we saw in Chapter 5, Umbriel and Oberon belong in the 'dead worlds' class. They are heavily cratered and few or no signs of internal activity have survived. Each of these moons has a 'twin', in terms of size and mass, occupying a neighbouring orbit of smaller radius. These are Ariel and Titania, and they bear a striking resemblance to one another rather than to their respective 'twins'. We will begin with Ariel, which was imaged at the higher resolution, *Voyager-2*'s closest approach to Ariel being three times nearer than its closest approach to Titania.

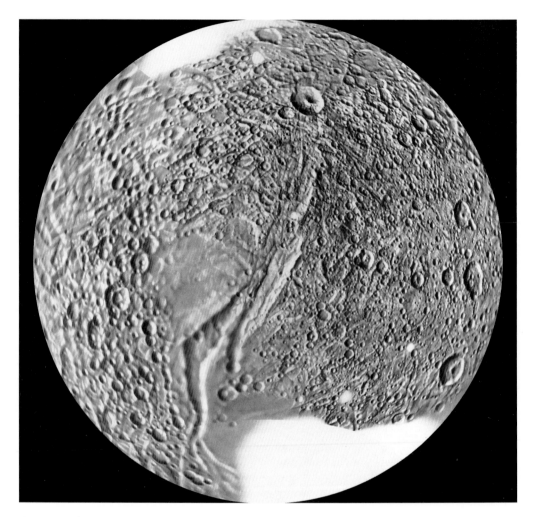

Fig. 6.20. Shaded relief maps of Tethys, showing the Saturn-facing hemisphere (*left*) and anti-Saturn hemisphere (*right*). Ithaca Chasma runs roughly north–south through the Saturn-facing hemisphere, and a trace of it can be seen in the north-west of the anti-Saturn hemisphere. The relief of the giant crater Odysseus is shown in an exaggerated fashion (compare with Fig. 6.19).

6.4.1 Terrain types and tectonic features

The best *Voyager* view of Ariel is shown in Fig. 6.21. This reveals a surface occupied by three distinct units, as summarized in Fig. 6.22: cratered terrain, ridged terrain, and plains. The cratered terrain is evidently the oldest, but it has a lower crater density than its equivalent on any other satellite of Uranus (Fig. 5.24). Almost all the craters are attribut-

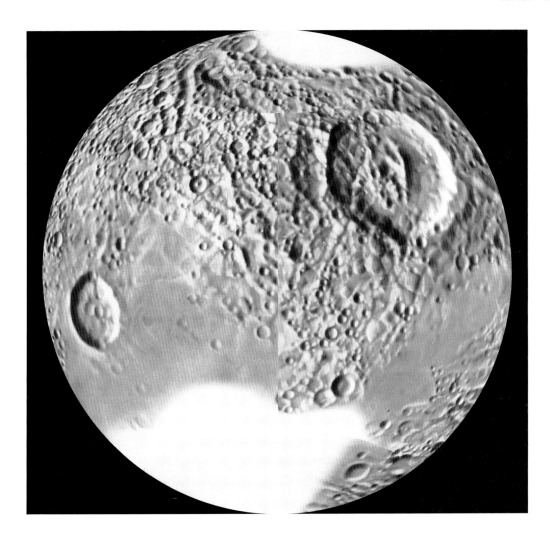

able to population II, although there are a few degraded and relaxed craters in the 50–100 km diameter range that may have survived from population I times. Essentially though, the cratered terrain appears to have been resurfaced after the population I bombardment had ceased, and, within the limits of the data, there are no significant differences in age within this unit. It has completely replaced any ancient heavily cratered surface that Ariel may originally have had, such as that preserved on Umbriel and Oberon.

Areas of the cratered terrain are separated by tracts of ridged terrain, which is characterized by parallel low ridges about 10–35 km apart with shallow troughs between them. The crater density on the ridged terrain is indistinguishable from that on the cratered terrain, and so the two units evidently date from the same era. Some of the ridged terrain belts are continuous with the major tectonic features of Ariel, a widespread system

Fig. 6.21. A mozaic made from the four highest resolution *Voyager-2* images of Ariel.

of faults that usually bound steep-sided troughs (chasmata) from 15 to 50 km in width. These are usually interpreted as grabens. Many of the chasmata are filled with a relatively smooth plains-forming material, which is the third of Ariel's terrain units. This plains unit provides the clearest evidence of fluid volcanism encountered so far in this book, where it has spilled out beyond the end of a chasma and flooded half of a 30 km diameter crater (Fig. 6.23). Where it is confined within a chasma, the plains unit commonly has a linear to sinuous medial groove in the form of a trough 2–3 km wide. The possible significance of these troughs will be discussed shortly.

The plains unit is markedly cratered, and it evidently dates back a long time. More significantly, the crater density on it varies from place to place in a manner that suggests that the plains-forming activity continued over a substantial period of time. This contention is backed up by the fact that some of the fault scarps have abundant craters

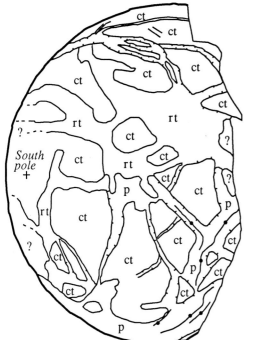

| Fault scarp
Groove

Fig. 6.22. Sketch map of the region of Ariel covered in Fig. 6.21 showing the major terrain units and tectonic features. The terrain units are: *ct*, cratered terrain; *rt*, ridged terrain; and *p*, plains.

superposed on them, whereas others (evidently the youngest) have far fewer craters cutting them (Fig. 6.24). The youngest features on Ariel appear to be several craters with bright ejecta blankets, in marked contrast to Umbriel, which, as we have seen, appears to have a uniform coat of darker material.

6.4.2 Tectonic processes on Ariel

It has been suggested that the overall pattern of linear tectonic features on Ariel represents fractures created by tidal forces, either by tidal despinning when its rotation became synchronous, in the same way as the linear features on Dione can be accounted for, or by collapse of a tidal bulge when the orbits of Ariel and Umbriel become non-resonant (if indeed they ever were in resonance: the present ratio of their orbital periods is slightly less than 5:3). How the fracture pattern on Ariel developed its present expression, irrespective of whether or not the basic pattern reflects a tidal origin, is a matter of considerable dispute, not least because only 35 per cent of the globe was imaged by *Voyager-2*. The tension indicated by the graben interpretation of the chasmata requires global expansion, unless there is a compensating amount of compression hidden in the un-imaged 65 per cent of the globe. Ariel is too small for internal ice phase changes, and so the most likely mechanism is the expansion that would have accompanied freezing of either water or an

Fig. 6.23. Detail from within Fig. 6.21 showing the plains unit spilling out beyond the end of Sylph Chasma and half flooding an older crater, Agape, slightly above right of centre.

ammonia hydrate melt in the interior. This would have had a much greater effect on the radius of Ariel than subsequent thermal contraction of the ice.

When looked at in detail, the grabens (if such they are) on Ariel present a rather unusual aspect. The blocks of high ground separated by these grabens sometimes meet at corners, so that the direction and amount of downthrow on some of the bounding faults change instantaneously from about 2 km in one direction to about 2 km in the opposite direction. Analogues for behaviour of this kind in extensional terrains on the Earth and elsewhere are rare. One way to overcome this problem is to allow sideways fault movement along some of the chasmata. Figure 6.25 shows how about 50 km of lateral motion across Korrigan Chasma and Kewpie Chasma could have resulted in the corner-to-corner situation and would at the same time account for the apparent offset and deformation of an older structure, Kra Chasma. Korrigan Chasma and the offset portions of Kra Chasma are visible on the extreme right-hand edge of Fig. 6.23.

6.4.3 *Volcanic processes on Ariel*

Whatever their tectonic origin and history, many of the chasmata on Ariel have clearly been flooded by some kind of volcanic flow process. The area imaged in Fig. 6.23 provides

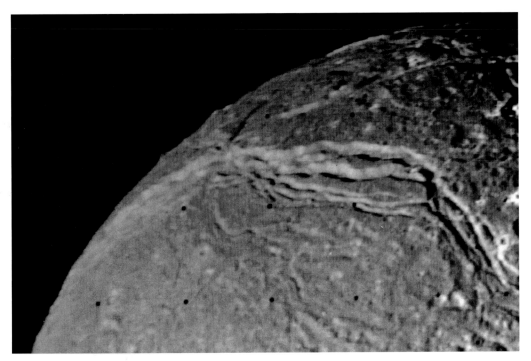

Fig. 6.24. Detail from within Fig. 6.21 showing an oblique view of the relatively pristine horst and graben structure of Kachina Chasmata.

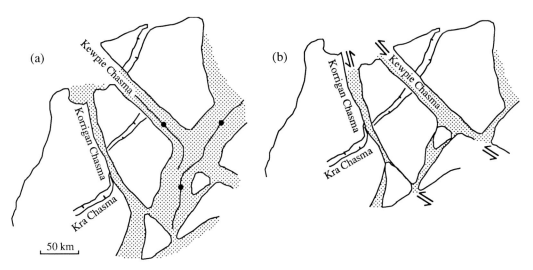

Fig. 6.25. Possible strike-slip faulting history of Korrigan Chasma and Kewpie Chasma. (a) Detail from Fig. 6.22 showing the present day situation of blocks of cratered terrain separated by younger chasmata (dotted areas). An older chasma, Kra, appears to have been offset by sideways (strike-slip) motion across Kewpie Chasma and Korrigan Chasma, and two blocks meet, uncharacteristically, at their corners. (b) Proposed reconstruction before the strike-slip faulting event, showing that Kra Chasma may originally have had no offsets. The double arrows show the sense of motion required to bring about the present day situation.

pretty much indisputable evidence of this. However, the nature of the flow material remains debatable. The range of possible models is reviewed below.

To test their hypotheses, various workers have tried to estimate the viscosity of the flow material on Ariel by measuring profiles across a flow. In the absence of stereoscopic images, the only way to do this is by photoclinometry, that is by measuring a brightness profile and using a model of how the surface material scatters sunlight at different angles to calculate the slope. We have to assume (not always justifiably) that there are no changes in either albedo or light-scattering properties along the profile. A typical photoclinometric profile across a flow in a chasma on Ariel is shown in Fig. 6.26. This shows that the flow has a strikingly convex top and that the edge of the flow is steep. The result is that the floor in the centre of the chasma approaches the height of the terrain on either side and the floor is deepest at its edges, next to the walls.

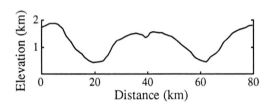

Fig. 6.26. Generalized profile across a flow-filled chasma on Ariel, as derived by photoclinometry. Note the vertical exaggeration of 10:1. The example shown contains a medial groove.

One school of thought takes such profiles at face value and regards them as revealing the original morphology of the flow, in an effectively unmodified state. Working on this assumption, the steepness of the edges of the flow can be used to calculate its viscosity, which works out at about 10^{16} poise virtually irrespective of the rate at which the flow was extruded. A viscosity of this magnitude is found in water-ice at a temperature of about 240 K; this is a much lower viscosity than that of lithospheric ice in the outer solar system generally, but it nevertheless represents an extremely viscous slow-moving flow, orders of magnitude more viscous than the most viscous lavas (rhyolite) on Earth. The steep edges of the profiles would represent the fronts of the flows, because these flows would have been far too viscous to have travelled along the length of a chasma, and would have to have been erupted from fissures that ran along the chasmata floors.

There are various problems with this approach. The most fundamental from a geo-logical point of view are how could such a high temperature (170 K above the ambient surface temperature of 70 K) be sustained near Ariel's surface, and how could the heat flow from this be localized to avoid viscous relaxation of nearby craters? For example, it has been calculated that a crater of 10 km diameter would relax completely over a time-scale of 100 years if heated to around 240 K. Thus the implied temperature makes pure water-ice an unlikely candidate for the flow material. However, it is possible to envisage other materials that would have the calculated viscosity at a lower, more feasible temperature, notably water-ice contaminated by volatiles such as ammonia, methane, nitrogen, or carbon monoxide. Traces of these in the flow material could act as lubricants for the creep of fine-grained crystalline ice at temperatures well below the 240 K required for pure ice, and this would make thermal relaxation of adjacent craters unlikely unless their ice had become similarly enriched in volatiles. On this model the sub-surface ice could be mobilized by the upward migration of 'warm' volatile-rich fluids, leading to extrusion of the flows through fissures parallel to (but not coincident with) the walls of the grabens. The flow-front advanced only short distances, say 10 km at most, before coming to rest. Where a medial groove occurs within a chasma, near the crest of the upwarp in the floor, it is interpreted (rather unsatisfactorily) as a junction between two flows.

An opposing school of thought notes that those viscosity models which suggest a value of 10^{16} poise are rather naïve, because they ignore the fact that the fluid may, like most geological materials, have possessed a yield strength. This would mean that its viscosity would appear higher at low rates of shear strain, so it ought not to be modelled as a simple newtonian fluid. Moreover, a low-viscosity fluid can produce a flow of apparently high viscosity if it is confined by a chilled, more rigid outer layer, or by an apron of debris shed from the advancing flow-front. When used to model terrestrial lava flows comparable models commonly give viscosities that are four to six orders of magnitude too high, and so it can be reasonably claimed that Ariel's photoclinometrically derived flow of profiles are compatible with flows of much lower viscosity. The most likely candidate would be an $NH_3 \cdot 2H_2O$ melt at 176 K (see Chapter 4) that had largely congealed so it consisted of more than 80 per cent crystals. A crystal mush like this could have a viscosity in the range 10^4–10^8 poise, which, allowing for Ariel's lower gravity, would behave much like a rhyolite lava flow on Earth. Similar behaviour would be found if the ammonia–water mix were contaminated with a trace amount of methanol, which would lower the freezing point by about 20 K and would keep the melt viscous even in the absence of any crystals. These are attractive mechanisms in that they avoid the high heat flow necessary to mobilize pure water-ice. Medial grooves can be explained by the coalescence of two flows, as before, or as the actual fissures from which the flows were erupted.

There are also ways to account for the medial grooves as faults or fractures. The simplest tectonic explanation is that they merely demonstrate an episode of renewed faulting on the trend established in the grabens, although it is difficult to see why this should occur along new lines instead of re-using the graben-bounding faults. A refinement of this concept is to suggest that the chasma floor has been bowed upwards after flow emplacement (either by viscous relaxation as in Fig. 6.4 or because of intrusion beneath the floor of the chasma) and that the medial grooves are fractures caused by stretching over the centre of the arch. However, if we appeal to non-volcanic processes to explain the profiles across the grabens, the estimates of flow viscosity based on these profiles become meaningless, except as crude upper limits. We ought also to be wary of the profiles themselves, because we have no good grounds for assuming that the albedo and particularly the light-scattering properties of the flow material are uniform and the same as those of the older surfaces to either side.

Having accepted that the profiles across a chasma may not tell us anything about the initial steepness of the flow margins, we can consider models at the opposite extreme of viscosity from where we started. What if the flows were erupted as crystal-free melts of $NH_3 \cdot 2H_2O$? Such a fluid would have a viscosity of the order of 100 poise (about the same

as honey) if crystal-free at 176 K, increasing by several orders of magnitude as crystallization progressed. While crystal-free, under Ariel's low gravity this melt would behave in a
slightly less mobile fashion than a molten basalt on Earth or the Moon, but would be much
more fluid than a terrestrial flow of andesitic composition. This opens up the possibility of
the flows spreading much further from their sources, so that a flow may have travelled
along the confining chasma from a point source at one end instead of oozing out sideways
from a fissure running the length of the flooded region.

The images we have are not adequate to show details of any of the putative source
regions on Ariel, but this mechanism is consistent with many of the observations that can
be made. In the first place, it is the most satisfactory way of accounting for the way that the
flows that extend beyond the ends of both Sylph and Korrigan Chasmata fan out as they
escape beyond the confining walls (Fig. 6.23); a chilled crust or rubble layer on the top and
sides of a flow would account for the steepness of its margins, whereas the low viscosity of
the interior of the flow would have enabled it to spread a long way. Secondly, it offers an
attractive way of explaining the medial grooves as collapsed lava tubes. Tubes commonly
form in basalt flows that travel many kilometres on Earth and hundreds of kilometres on
the Moon. The chilled roof of the tube acts as an insulating cap that prevents significant
cooling of the melt, enabling it to travel great distances before solidifying. The roof of a
tube that later becomes drained of melt is liable to collapse, and this is the favoured origin
for most of the sinuous rilles on the Moon, some of which bear a striking resemblance to

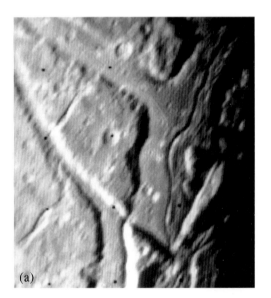 

Fig. 6.27. (a) The best *Voyager* view of Sprite Vallis, a 2–3 km wide medial groove in Brownie
Chasma. (b) A *Lunar Orbiter* mozaic showing the sinuous rille in the lava-flooded Alpine Valley on the Moon. These two examples have comparable scales, and lend support to the concept
that medial grooves on Ariel may be collapsed cryovolcanic 'lava' tubes, although there are
several other possible interpretations.

Ariel's medial grooves (Fig. 6.27). There are a few examples on Ariel where a medial ridge takes the place of a medial groove. These can be interpreted as tubes that did not drain and where contraction or subsidence of the rest of the flow surface left the tube as a positive feature.

Perhaps it is unwise to seek a single model for the viscosity and composition of the flows on Ariel. Debates of this nature are likely to continue until such time as we have higher-resolution images covering more of Ariel's surface, although our appreciation of what might have occurred will undoubtedly improve as we learn more about the composition of the outer solar system. Furthermore, the effects of crystallinity and trace contaminants on the viscosity of low temperature melts are only just beginning to be documented. It is probably being unreasonably optimistic to expect real flows to be pure $NH_3 \cdot 2H_2O$, and we badly need to know more about what happens when methane, methanol, and other volatiles are present in the melt. These matters are being investigated by laboratory studies of ice mixtures at low temperatures, and at both low and high pressures. As this work continues, the choice of possible explanations for Ariel's flows is likely to become greater rather than less.

6.4.4 *Geological history*

In view of the likely complexities of Ariel's history, and its comparatively volatile-rich composition, convection in the solid state, if not actually involving liquids, is likely to have occurred to a sufficient extent to permit almost complete differentiation, with the possible exception of an undifferentiated crust that retains its primordial rock–ice mixture except where covered by flows. The thermal evolution of Ariel may well have followed the same lines as that of Dione, which has a similar size, density, and surface temperature. The tensional features are best explained in the same way, that is as a result of expansion due to the onset of freezing of interior melts. We have noted that the global tectonic pattern of Ariel can be interpreted as being due to tidal stresses. If Ariel was at times in orbital resonance with Umbriel, this would have resulted in one or more episodes of renewed interior heating, possibly involving even the melting of water, the re-freezing of which would have encouraged graben formation. However, it is not certain that the relative motion across the chasmata was dominantly extensional in all instances, and some features are more elegantly interpreted as being due to lateral displacements of the order of 50 km.

Tidal heating aside, Ariel's declining rate of radiogenic heat production was probably sufficient to liberate a melt of $NH_3 \cdot 2H_2O$ composition over about the first two billion years, and methane would have been mobile for even longer. The eruption of such melts would have been more likely in the earlier stages, when they could have been generated closest to the surface. Later on, the major flows are likely to have escaped to the surface along fractures that opened up under tension, hence the intimate association between the flows and the chasmata. We shall see further evidence of eruptions and their relationship with fault systems on the remaining two major moons of Uranus; Titania and Miranda.

6.5 Titania

Titania is comparable in size and mass to Oberon, although in appearance it is more similar to Ariel. Unfortunately, the best *Voyager* images have a resolution of about 7 km, so we cannot compare any details. An airbrush map of Titania's southern hemisphere is shown in Fig. 6.28. Titania has abundant population II craters, with a few 100–200 km basins that may be relics of population I craters. The most cratered regions have a greater

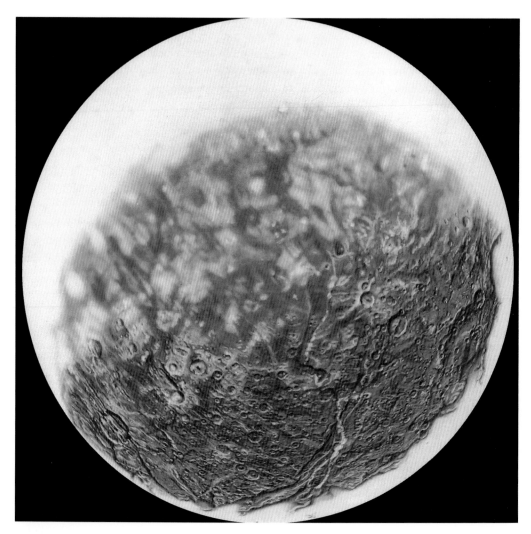

Fig. 6.28. Shaded relief map of Titania's southern hemisphere. The resolution of the *Voyager-2* images was poor except in the lower (Uranus-facing) half. The faulted terrain of Messina Chasmata that extends upwards from the lower edge is reminiscent of Kachina Chasmata on Ariel (Fig. 6.24).

crater density than the equivalent regions on Ariel (Fig. 5.24), which may mean that Titania has terrains surviving that are older than anything seen on Ariel. The biggest craters appear to have a somewhat relaxed topography. There are several patches with markedly lower crater densities, suggesting resurfacing probably by volcanic processes, but possibly by viscous relaxation in response to localized heating. The surface is transected by major faults forming scarps of between 2 and 5 km in height, and as much as 1500 km long. Opposite-facing fault scarps commonly occur in pairs or multiple sets forming 20–50 km wide grabens or horst-and-graben belts in a way strongly reminiscent of Ariel, although the temporal and spatial relationships between volcanic flooding and the grabens are less clear. The faults must be among the youngest features on Titania, because they can be seen to cut through craters and rarely have younger craters super-imposed on them. Like Oberon, Titania's youngest craters are marked by bright ejecta blankets, indicating either a cleaner ice layer below the surface, or a relationship between exposure age and radiation-darkening. The same phenomena are indicated by the bright-ness of the sunward-facing walls of the grabens, which is unlikely to be due entirely to the lighting effect.

The faulting on Titania is most simply explained as a result of an episode of global expansion, which can be attributed to internal freezing in the same manner as the grabens on Ariel and Dione. If the resolution of the images were better, Titania might well show clearer evidence of volcanic processes, but at present we cannot be sure whether or not any of the chasmata have been flooded in ways comparable to the chasmata on Ariel. However, the implications of what we can see are that Titania has been more active than its neighbours on either side, Umbriel and Oberon, and it is probably necessary to appeal to an episode of tidal heating at some time in the past to explain this activity.

6.6 Miranda

Miranda is the character in *The Tempest* who speaks the words 'O brave new world'. William Shakespeare was undoubtedly a genius in his own field, but he could surely not have foreseen that one of the most remarkable worlds in the solar system (Plate 1) would come to be named after his character.

6.6.1 *Terrain units*

It was a happy chance that led to Miranda being among the best imaged bodies (in terms of resolution) in the outer solar system. *Voyager-2*'s trajectory through the Uranus system (Fig. 3.7a) was chosen to take it close to Miranda in preference to any of the larger satellites simply because it was the innermost satellite known when the mission was being planned, and the spacecraft needed to pass close to Uranus to gain a gravitational assist onwards towards Neptune. The mozaic of high resolution images in Fig. 6.29a reveals a startlingly complex surface. About half of it has a relatively uniform albedo and a rolling topography, densely pocked by craters down to the limit of resolution. This is the most

(a)

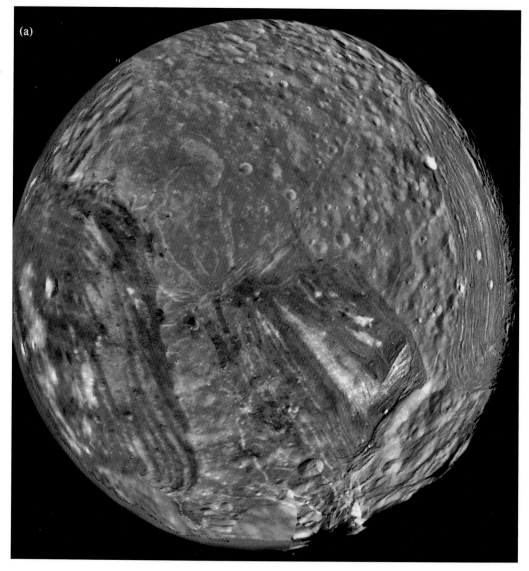

(b)

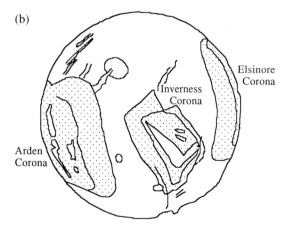

Elsinore
Corona

Inverness
Corona

Arden
Corona

Fig. 6.29. (a) Mosaic of *Voyager-2* images covering the hemisphere of Miranda that was sunlit at the time of the encounter. The Uranus-facing hemisphere is towards the bottom. (b) Sketch maps of terrains, same orientation as (a).

densely cratered surface of all the Uranian satellites (Fig. 5.24). A large proportion of these craters have a 'softened' appearance about their rims; this is unlike viscous relaxation and is probably due to the deposition of a widespread mantling deposit of some kind. The rest of the craters are much fresher-looking (there are very few with transitional morphology) and commonly reveal a bright layer many hundreds of metres thick in their walls. The same layer can sometimes be picked out in fault scarps, and this is probably the material that forms the mantling deposit over the older craters. The other half of the surface is occupied by three regions of younger terrain of a nature not encountered elsewhere in the solar system, the interiors of which are marked by belts of bright and dark albedo material and whose margins are occupied by belts of parallel ridges. The crater density here is lower than anywhere else in the Uranian system, and the craters are all fresh in appearance. These regions have been termed *coronae* (Fig. 6.29b), and each displays somewhat different characteristics.

In keeping with the Shakespearian motif, the coronae are named after places that feature in his plays. The corona at the right of Fig. 6.29 lies in the trailing hemisphere, and is named Elsinore Corona (after the setting for *Hamlet*). This appears to be the least complex of the three, although much of it is out of sight. A detail of part of it is shown in Fig. 6.30. The inner part, which is seen on the terminator, is a chaotic region of presumably fault-bounded blocks, without the major albedo differences found in the other coronae (possibly an effect of the low angle of solar illumination in this region). This is surrounded by a belt of parallel ridges, each up to 1 km high and several tens of kilometres long, the outer part of which appears to be the youngest because it turns the corners at the ends of the corona (Figs 6.30 and 6.32), thus truncating the section of ridges interior to it.

The ridges are most commonly interpreted as fracture-controlled extrusions of warm ice or a melt from among the high-viscosity options suggested for the flows on Ariel, although they might be sites of shallow intrusion, and some may reflect fault motion. The outermost ridges in places appear to be developed on top of the cratered terrain, suggesting that the ridge-forming process propagated outwards from the centre of the corona. The youngest feature of all in this region occurs near the top left of Fig. 6.30 and bears a striking similarity in morphology to a terrestrial dacite lava flow, being clearly less viscous than the material forming the ridges if they are flows as well. A sketch interpretation is shown in Fig. 6.31.

The cornona near the centre of Fig. 6.29 is named Inverness Corona (after the site of Macbeth's castle). It has a ridge belt around its margin comparable to that of Elsinore Corona, but its centre is rather different. There is a chevron-shaped pale albedo pattern, the outline of which is paralleled by a ridge and furrow topographic grain within it, that stops abruptly against the outer ridged belt (Fig. 6.32). The whole structure is cross-cut by younger normal faults that run on into the adjacent cratered terrain, one of which becomes dominant by taking over the vertical motion of several smaller faults to result in a scarp over 10 km high (Fig. 6.33).

The third corona, Arden Corona (after the setting for *As You Like It*), lies on the left of Fig. 6.29, in Miranda's leading hemisphere. It has a well-developed belt of parallel ridges around part of its periphery, like Elsinore Corona except that the albedo is non-uniform

Fig. 6.30. Part of Elsinore Corona, with cratered terrain towards the left. Note that the craters within the corona are all sharp; there is a similar density of sharp craters in the adjacent cratered terrain, but there are also many older craters with softened edges that may have been mantled by a global event. Image about 200 km across.

and there are dark strips parallel to the ridges. The central part of Arden Corona has bright albedo features reminiscent of Inverness Corona, although they do not form such a regular pattern. The view looking across the limb of Miranda in Plate 1 shows detail of the edge of Arden Corona (from a part of the upper edge of the corona in Fig. 6.29, some way in from the edge of the globe). The tonal bands in the outer, ridged, zone stop abruptly where they come up against the cratered terrain, which is intensely faulted in this region. The contact with the cratered terrain makes a 90° turn here and as it runs over the limb there is an impression of rotated fault blocks within the outer edge of the corona, as interpreted in Fig. 6.34.

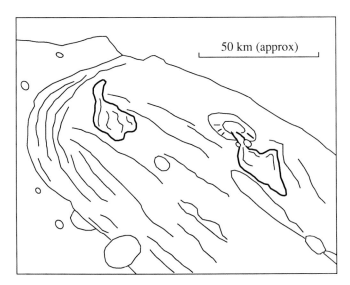

Fig. 6.31. Sketch map of the upper left part of Elsinore Corona in Fig. 6.30 showing two probable 'lava' flows (heavy outlines). The one on the left is the most obvious; the other is harder to make out, as it is partly hidden in the shadow of the ridge in the foreground.

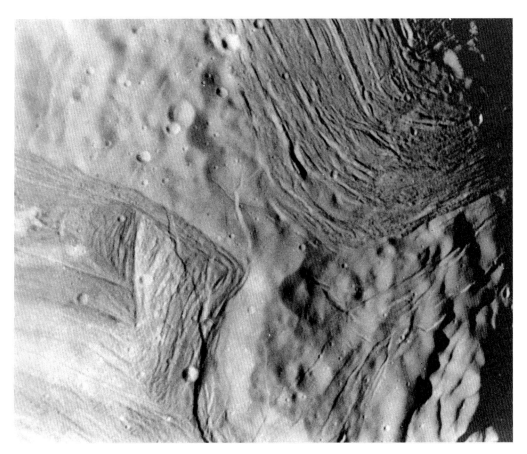

Fig. 6.32. Part of Inverness Corona with its pale chevron (lower left) and a corner of Elsinore Corona (upper right). Note the younger brittle fault structures especially in the lower part of the image. A bright deposit is exposed in the inner walls of the two fresh 7 km diameter pit craters between the two coronae, towards the upper left; this is probably the mantling deposit that softens the morphology of the older craters. Image 220 km across.

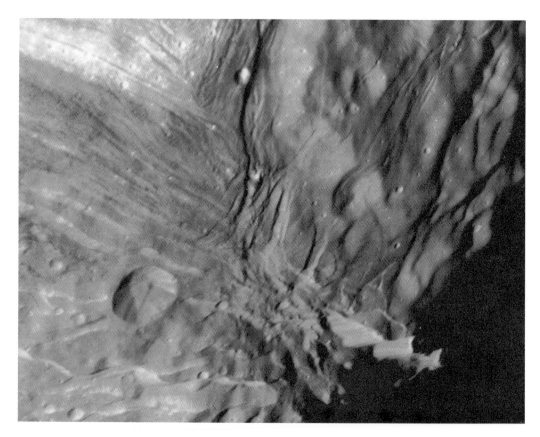

Fig. 6.33. Detail of fault scarps near Inverness Corona, from the lower part of Fig. 6.29a. The major scarp on the terminator has a slope of about 60° and is over 10 km high, and there are several lower scarps parallel to it to the left, which affect the cratered terrain and the corona. The fractures immediately behind the major scarp suggest that part of it is ready to collapse in a giant landslip, and the scalloped nature of the scarp as it disappears into darkness suggests this has already happened in places. If the terrain beyond the foot of the scarp were in sunlight, deposits formed by debris avalanches might be seen.

Fig. 6.34. Sketch cross-section of the edge of Arden Corona as seen in part of Plate 1, interpreted as a series of rotated fault blocks. There is darker material below the original surface, which is exposed on the rotational faults.

6.6.2 *Geological history*

Making sense of Miranda's history is one of the more formidable problems in planetary science. One of the earliest suggestions to explain the coronae, particularly Inverness Corona with its bright chevron sandwiched between dark belts, was that Miranda as we now see it has been re-accreted following collisional break-up, and that the tonal belts are layering from the earlier body exposed end-on within large, corona-sized, re-accreted fragments. Attractive as this idea is in view of the statistical arguments that gravitational focusing of impactors would have caused Miranda to have been disrupted in this way (page 85), it does not account for the dramatic difference in crater density between the coronae and the cratered terrain. Moreover, it is difficult to imagine how any such layering could have survived with so little distortion while the re-accreted Miranda became spherical again under its own gravity. On the other hand, how could such peculiar geometries as those shown by the coronae result from essentially volcanic processes?

A compromise model that can explain most aspects of Miranda's geology is outlined below. It suggests that the patterns in the coronae are produced by a combination of tectonics and tectonically controlled volcanism, but that the initiation of the coronae is a result of impacts during re-accretion.

The crucial observation is that numerous fractures run across the face of Miranda, varying from old and subdued to younger and fresher breaks. The latter cut the coronae as well as the cratered terrain. Many of the fractures are in parallel sets with opposite-facing scarps and are most obviously interpreted as horst-and-graben structures. These represent global expansion of, at most, a few per cent, presumably as a result of thermal expansion in the interior or of freezing of an internal molten zone in the manner suggested for Dione and Ariel. The fractures tend to run either north-north-west to east-south-east or east-north-east to west-south-west, with the same trends being reactivated several times during Miranda's history. These fracture directions are reflected in the general outlines and orientation of the ridges and tonal bands in Inverness and Arden Coronae, and to a lesser extent in Elsinore Corona; so whatever disturbances initiated the coronae, the surface manifestations were evidently controlled by the ancient fracture directions. The problem now becomes one of finding a mechanism for producing concentric structures like the coronae, taking advantage of this fracture pattern.

This is where the re-accretion idea can be used. Assuming that the version of Miranda that existed prior to its most recent collisional break-up was at least partly differentiated, there would have been some silicate-dominated fragments among the debris so produced. On re-accretion, rocky and icy fragments would have re-grouped randomly. The impact velocities would have been low, and accretional heating would have been far too low to allow the new Miranda to differentiate. Although radiogenic heating might have been capable of mobilizing the deep interior by partial melting of ammonia–water-ice, this probably did not lead to significant differentiation. If we wish to invoke near-surface melts, as implied by the evidence of volcanism (e.g. Fig. 6.31), then there probably needs to have been an external heat source, such as tidal heating during episodes of orbital resonance with Ariel or Umbriel. This then could give rise to a differentiated Miranda, with a thin icy lithosphere overlying a 'warm' icy asthenosphere and a rocky core. The late accretion of a silicate-rich fragment 10–20 km in diameter now comes into play to form a corona.

This fragment could either already have been accreted in the outer part of Miranda's body, or be a piece that actually arrived during the tidal heating event. It would arrive with low velocity (if it were derived from the break-up of the earlier Miranda), so it would not be severely disrupted upon impact. Instead, having penetrated the lithosphere, it would sink through the less dense icy asthenosphere. Figure 6.35 shows the type of flow pattern that this movement would cause. The stresses on the overlying lithosphere are compressional, and would result in chaotically oriented compressional features on the surface directly above the sinking mass, changing to concentric compressional features at distances greater than 15° from the centre. The central chaotic region of Elsinore Corona and its outer zone of concentric ridges (Fig. 6.30) indicate strongly that asthenospheric flow of this kind took place. The spacing of features is compatible with a lithospheric thickness in the range 3–30 km at the time of formation. Arden Corona could have formed in the same way, although it has also been suggested that it began as an impact basin generated by the incoming silicate fragment, and that this event caused the widespread distribution of the pale deposit that mantles the older craters.

The subsequent appearance of the coronae has evidently been affected by later fault movements (Fig. 6.34) and at least some volcanism (Fig. 6.31). As has already been remarked, the form of the ridges surrounding Elsinore Corona is compatible with their having been constructed by the extrusion of viscous fluids, but they could also represent linear swells above rising intrusions of less-dense material. Compressional stresses do not encourage the extrusion of melts, because such stresses tend to close up any pathways to the surface. Melting caused by the corona-initiating impact is one way to produce melts in a setting where they would have had easy access to the surface. Another is to appeal to later extensional forces re-activating the original compressional features of a corona. A third is to reject the compressional nature of the corona-forming tectonics by abandoning the dense 'sinker' model and appeal instead to a rising mass of less dense material, such as a silicate-free ice fragment segregating upwards from Miranda's centre. This would produce the same flow lines as in the 'sinker' model of Fig. 6.35, but the flow would be in

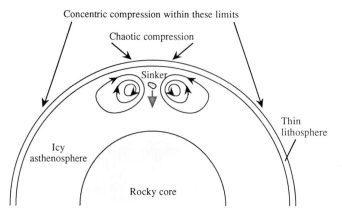

Fig. 6.35. Flow in a viscous asthenosphere caused by a sinking mass, showing the resulting stresses in the overlying lithosphere. This process could initiate the formation of a corona on Miranda.

the opposite direction. The surface breaks would have the same orientations as before, except that extension would take the place of compression.

It should now be apparent that there is no shortage of theories that could account for the coronae and various aspects of their morphology. What is still lacking is a general consensus about what really did happen. Inverness Corona is less clearly explicable on the 'sinker' model than either of its companions, and it can be accounted for entirely by volcanic processes controlled by the pre-existing fracture pattern, although why volcanism should be concentrated in this region remains unexplained. A geological interpretation is shown in Fig. 6.36. Four units are identified. On this model, unit A is stratigraphically the lowest and therefore the oldest unit in the corona. It is medium-dark and is characterized by ridges parallel to the outer edge. The edge itself is a convex lip that

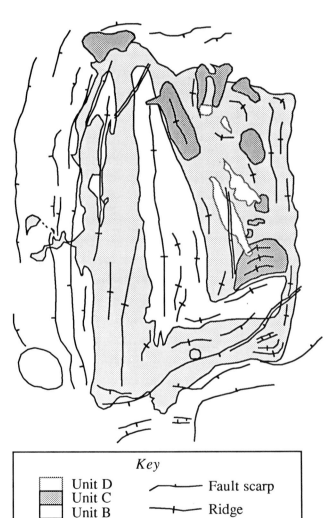

Fig. 6.36. Geological interpretation of Inverness Corona, dividing it into four units, A–D. See text for explanation.

Key

Unit D
Unit C
Unit B
Unit A

⌐—— Fault scarp

—+— Ridge

50 km

in places has outward projections into adjacent hollows. This is consistent with an origin as a flow deposit, which is made even more clear if some dark markings beyond the western edge of the corona are interpreted as extensions of this unit spilling over the fault scarps and ponding in craters and other hollows.

Unit B forms the bright chevron in the centre of Inverness Corona. It convex lip suggests that it overlies unit A. Apart from its higher albedo it is similar to much of unit A. It may represent a purer ice with fewer silicate impurities, or a flow of identical composition to unit A but with a different surface chilling history. Unit C overlies both A and B. It is slightly darker than unit A and has superimposed its own pattern of ridges over the pre-existing corona topography; it appears to represent similar material to unit A but fed from localized vents rather than through fissures parallel to the overall fracture pattern. Generally, the ridges in units A–C are compatible with a viscosity of around 10^8 poise, implying that the flows consisted of a liquid–crystal mixture, like one of the flow materials suggested for Ariel. Unit D consists of bright patches with no apparent topographic effect, and is probably a dusting of pyroclastic 'snow' from explosive eruptions or patches in the earlier units that became discoloured as a result of some form of fumarolic activity caused by escaping gases.

Interpretations such as the above, relying on observations made at the limits of the resolution of the images, are subject to considerable dispute. However, provided they are correct in essence, if not in detail, it is apparent that volcanism on Miranda at least, and presumably on certain other icy satellites, is comparable to that encountered on Earth in terms of its complexity and variety of morphological expression.

We have now completed our survey of the recently active moons, and can turn to those where geological activity is continuing today. We shall begin with Io, where there is a variety of features whose volcanic origin is in no doubt whatsoever.

Plate 1. Composite of *Voyager-2* images showing a close-up view of part of Miranda with a resolution of less than 1 km, silhouetted against the planet Uranus. Arden Corona lies in the left-hand region of Miranda, with cratered terrain to the right. An interpretation of the faulted structure of the edge of the corona is given in Fig. 6.34.

Plate 2. *Voyager-1* image of a characteristic part of Io's surface, in high southern latitudes, 1800 km across. There are many vents, notably the shield volcano Maasaw Patera (top centre) and below it the elongated floor of Creidne Patera partly covered by dark flows. Layered plains material occupies much of the image, especially near the centre, with the prominent mountain Haemus Mons jutting above it in the lower left corner.

Plate 3. *Voyager-1* view looking towards the limb of Io, showing Pele in eruption. The vent is in the rugged volcanic complex towards the upper left. The plume is visible only against the background of space, where it has been enhanced to make it show up.

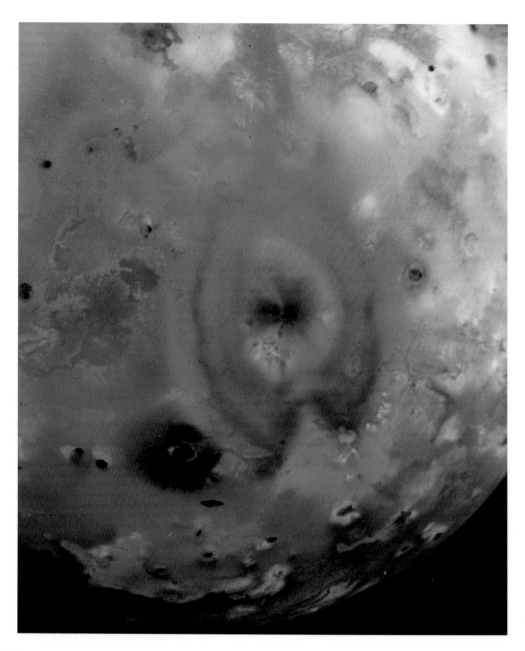

Plate 4. *Voyager-1* image looking down to the surface of Io through the Pele plume. The plume itself is virtually transparent, but the dark concentric deposits resulting from it can be clearly seen. By the time *Voyager-2* was able to image this area, the 'bite' missing from the southern rim of the deposit had been filled in, but the plume was no longer active. The image is about 1850 km across.

Plate 5. *Voyager-1* image covering a region of Io shown in the upper left of Fig. 7.2. Loki fissure is the dark linear feature adjacent to the larger dark patch of Loki Patera, near the centre of the view. Faint white plumes are erupting from either end of Loki fissure. Ra Patera (Fig. 7.12) is visible in the lower left. The dark spot to the north of Loki is Ameratsu Patera, which was recorded as a hot-spot by the *Voyager-1* IRIS instrument. In the white region near the limb to its left lies Surt, the probable site of a Pele-type eruption between the two *Voyager* encounters.

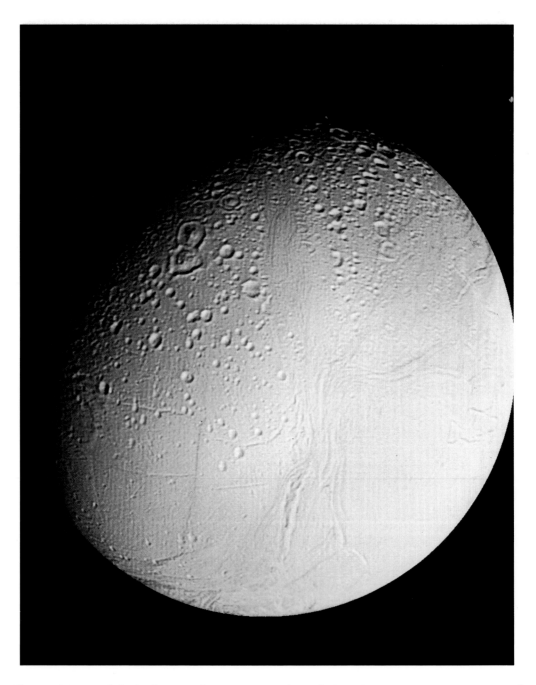

Plate 6. Mozaic of the highest resolution images of Enceladus, showing craters in a variety of states of degradation and considerable evidence of tectonism and resurfacing.

Plate 7. Mozaic of the highest resolution images of Triton. The equator runs more or less from top to bottom through the centre of the plate. The south polar cap of nitrogen-ice occupies the left-hand side, cantaloupe terrain occupies the upper right and below this are 'smooth' high plains (near the terminator) and hummocky terrain (between the smooth plains and the polar cap). A terrain map of the area covered by this mosaic is given in Fig. 7.29.

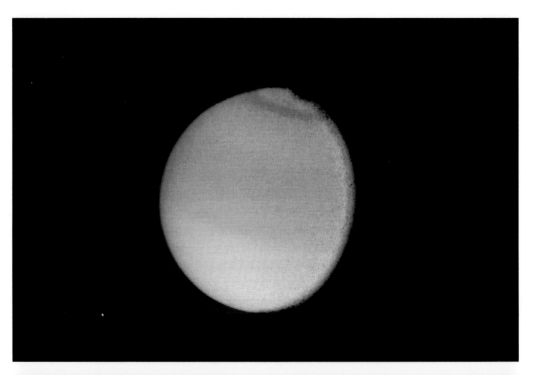

Plate 8. Two *Voyager-1* views of Titan. The red colour is due to traces of nitrogen-bearing organic molecules within the opaque aerosol layer, 200 km above the ground surface. Top: general view showing the contrast in brightness of the atmosphere between the southern and northern hemispheres and the dark north polar hood. Bottom: blue haze layers seen above the limb.

7

Active worlds

O brave new world . . .
Shakespeare, *The Tempest*

Four satellites are regarded as active in this book: Io and Europa (Jupiter), Enceladus (Saturn), and Triton (Neptune). The occurrence of present day activity on Europa and Enceladus can be disputed, but there is no doubt whatsoever in the case of Io.

7.1 Io

Unlike all the other satellites discussed in this book, Io is not icy. Its density and size are intermediate between those of the Moon and Mars, and in many ways it can be regarded as a terrestrial planet that happens to be in orbit around Jupiter. Jupiter's influence is manifested by the elevated temperature in the inner part of the proto-jovian nebula that seems to have prevented a substantial amount of ice from accreting into Io. Orbital resonance with Europa keeps Io's interior hotter than it would otherwise now be in a rocky body of this size.

If interest by geologists in the moons of the outer planets can be said to have been sparked by a single event, this event would be the discovery of active volcanism on Io. As noted in Chapter 1, the prediction that tidal heating would maintain Io's interior at a high temperature was dramatically borne out when eruption plumes were discovered on some of the *Voyager-1* images (Fig. 1.7).

The clues were already there, however, that Io is an unusual place. Some of the earliest Earth-based photometric observations had shown that Io is the reddest known object in the solar system, with the possible exception of Mars, and it became progressively more certain during the 1970s that Io lacks the ice that can be detected in the reflectance spectra of other satellites. The red colour and a particular absorption feature in the ultraviolet led some workers to suggest the dominance of sulphur on Io's surface. In 1964 it was noticed that bursts of radio waves from Jupiter were correlated with Io's orbital position. Gradually, the strong interaction between Io and Jupiter's magnetic field was revealed, culminating in the 1973 fly-by of *Pioneer-10* that proved that Io's orbit actually lies within the planet's magnetosphere. The discovery in the same year of an emission line due to sodium in Io's spectrum was the first step on the road to our present knowledge that Io is surrounded by a cloud of neutral sodium, potassium, and oxygen atoms, dispersed fore and aft of Io approximately along its orbit. Shortly afterwards, ionized species, notably sulphur and oxygen, were discovered in Io's vicinity, and we now know from *Voyager* data that these are confined by Jupiter's magnetic field into a doughnut-shaped torus that runs right round Jupiter at the radius of Io's orbit (Fig. 7.1). It is now clear that the sulphur and other atoms forming the torus and cloud are supplied from Io's interior by volcanism, at least partly by way of plumes such as those detected on the *Voyager* images.

Another clue came from telescopic thermal infra-red measurements; these show that Io's surface temperature is about 130 K by day, which is a reasonable value in this part of the solar system. However, it was discovered during the 1970s that when Io passes into the shadow of Jupiter, the amount by which its temperature drops appears less when measured at shorter infra-red wavelengths than at longer wavelengths. At first there was no acceptable explanation of this phenomenon, but it can now be explained by the presence of small volcanic hot-spots on Io's surface whose effective temperature is typically above 300 K. These make only a small contribution to the measured infra-red flux from Io when it is in sunlight, but become significant when the sunlight is cut off. The peak radiation from these hot-spots is in the short-wavelength infra-red, so when Io's temperature is calculated by measuring the infra-red radiation from the whole of its disc the value appears higher at progressively shorter wavelengths.

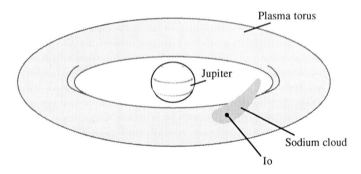

Fig. 7.1. Plasma torus and neutral sodium cloud associated with Io. The cloud lies in Io's orbital plane, and the leading part of it is skewed inwards towards Jupiter.

7.1.1 *The surface of Io*

Voyager-1 flew by Io at a distance of 20 500 km, an even closer encounter than *Voyager-2*'s fly-by of Miranda. The most detailed images have a resolution of slightly less than half a kilometre. These cover only limited areas, and a smaller proportion of the surface was imaged with a resolution of less than 2 km than for either Ganymede or Callisto (Fig. 3.6). Nevertheless, the data are adquate to show convincingly that impact craters are absent on Io. For the first time in this book we have reached a world where the surface is sculpted entirely by geological forces from within (Fig. 7.2 and Plates 2–5). In this case the morphology and surface markings appear to be very largely the result of volcanic processes. The redness of the surface suggests that sulphur, or oxides of sulphur, is important at least as a colouring agent on Io, although the relative importance of sulphur volcanism versus silicate volcanism in constructing the surface remains controversial. We

Fig. 7.2. Shaded relief map of Io's trailing hemisphere, showing a surface dominated by the action of volcanic processes. The leading hemisphere appears similar, but was imaged at poorer resolution.

shall consider some of the evidence shortly, but first we must describe the general nature of Io's surface.

The wide diversity of features on Io can be broken down into three categories: vent regions, plains, and mountains. Examples of all three types are well seen in Plate 2. In recognition of the discovery of active volcanism on Io, each of the vents is named after a mythological figure associated with fire or thunder. Some vent regions consist of a gently sloping volcano with a steep-walled, flat-floored summit crater, or caldera, several tens of kilometres in diameter. Radiating away from the summits there are usually dark markings that have every appearance of being lava flows of some kind. Maasaw Patera is a good example of this type of feature (Fig. 7.3 and Plate 2). It is strongly reminiscent of terrestrial basaltic shield volcanoes such as in the Hawaiian and Galapagos Islands (Fig. 7.4), although whether the lava flows on Io occur as molten silicate rock or as molten sulphur is a matter of debate.

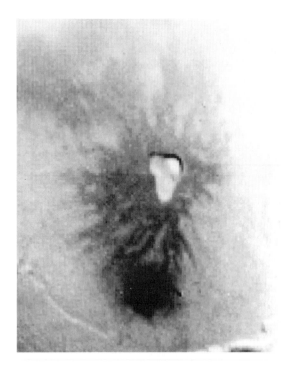

Fig. 7.3. Detail of Maasaw Patera, a shield volcano on Io. The summit caldera is over 2 km deep, and the image covers just over 100 km across.

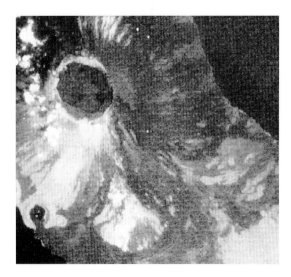

Fig. 7.4. A satellite image of Volcan Darwin, a basaltic shield volcano in the Galapagos Islands, which is morphologically similar to Maasaw Patera (Fig. 7.3) and other shield volcanoes on Io. This image is about 35 km across. The dark areas at top right and bottom left are the sea.

This sort of vent is gradational into large calderas up to about 200 km in diameter that are inset into the surface of the plains, without appearing to be at the summit of any edifice, except where lava flows can be discerned that have flowed away from the rim. The inner walls of the calderas are steep and scalloped, and the floor may occur in several levels. Such calderas (albeit these are on a larger scale, and generally less round) are reminiscent of terrestrial examples that are formed by collapse of the floor after a body of magma has been erupted by means of lava flows or in an explosive (pyroclastic) event. The floors of many of Io's calderas are wholly or partly covered by dark material, which is taken to be fresh lava. Examples are shown in Fig. 7.5.

In addition to the common types of volcanoes just described, Io has two known examples of a rather remarkable kind of volcanic construct, shown in Fig. 7.6. These are circular mounds about 180 km across, each with a summit crater and bounded by an apparently steep scarp running round their bases. In general form they bear at least a superficial resemblance to Olympus Mons on Mars, the largest volcano in the solar

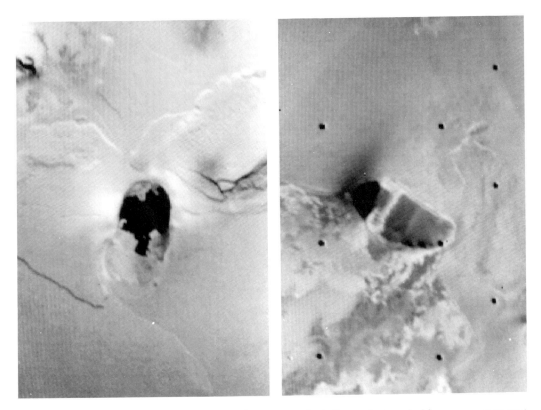

Fig. 7.5. Two paterae on Io, each with dark areas on their floors that probably represent recent lava of either silicate or sulphur composition. *Left*: Creidne Patera, 160 km from north to south, flanked on either side by layered deposits of unknown origin. Creidne Patera is also visible near the centre of Plate 2. *Right*: Tol-ava Patera, 80 km in length, with blotchy areas to its south suggestive of lava flows. The black dots are reseau marks.

Fig. 7.6. Two adjacent circular volcanic constructs on Io, Apis Tholus to the north and Inachus Tholus to the south, seen foreshortened in this image. Each is about 100 km across. Their origin is enigmatic. The dark feature to their south is another example of the more common type of Io volcano, Masaya Patera.

system, whose basal scarp is largely the result of landslides. No individual flows can be made out on the flanks of the examples on Io, suggesting that they were constructed by flows of a particularly fluid nature that spread symmetrically away from the vent.

Of the nine eruption plumes detected by *Voyagers-1* and *-2* (of which more later), each had its source at or near a vent of one of the sorts just described; six of them actually within calderas and three from dark linear fissures adjacent to a caldera. Furthermore, the Infrared Interferometer Spectrometer (IRIS) carried by *Voyager-1* observed three of the plume sources, demonstrating that they had anomalously high temperatures, and it revealed seven other hot-spots where there was apparently no plume at the time. The spatial resolution of the IRIS instrument was low, with a footprint 70 km or more across depending on range, so locations are approximate; however, each of the non-plume hot-spots was coincident with a caldera (one was at Creidne Patera, a caldera shown in Fig. 7.5) or a flow emanating from a caldera. Clearly many of Io's vents are still active, and apart from the plume and hot-spot evidence it is common to find pale discoloured areas, possibly condensed sulphur dioxide, around either a whole caldera or just an individual flow that provides evidence of recent fumarolic activity.

Most of Io's surface is covered by plains areas with little apparent relief. The plains are often blotchy in appearance with various light and dark patches. Such colour changes have been variously attributed to lava flows of different ages or different compositions, deposits formed by condensation of gases emitted at fumaroles, and chemical alteration of the ground due to the passage of such fumarolic gases. In regions where the Sun was low enough in the sky to cast shadows, some areas can be seen to consist of superimposed layers forming tabular, smooth-topped plateaux bounded by escarpments several hundreds of metres in height. Much of the central and southern parts of the area shown in Plate 2 is of this nature. On Earth, similar morphology would be explained by flat-lying layered deposits being eroded away at their edges. The extent of the layers is greater than that of the lobate lava flows that can be mapped emanating from Io's vents, so if the layers are formed by flows these must be of a different kind. Fall-out from eruption plumes is a possible cause of the layering; the thickness deposited at any one time ought to vary with distance from the vent, but if the force of the eruption were to change over time the thickness of the deposit as a whole could average out to a more uniform value. An alternative explanation of the layers is that they are analogous to terrestrial ignimbrite sheets, which are formed by ground-hugging pyroclastic eruptions arising from major calderas (Fig. 7.7). However, it is unclear whether on Io, in the absence of an appreciable atmosphere, the amount of gas necessary to sustain such a flow could be released by the eruption processes and remain entrained within the flow for long enough. Yet another explanation could be that the layered deposits are the result of gravity-driven flows descending from the high-standing mountain blocks (described later).

Fig. 7.7. Terrestrial ignimbrite flow in northern Chile, which was erupted from an explosive silicic caldera. The area shown is about 18 km across. Compare this with the tabular plateau to the northeast of Creidne Patera in Fig. 7.5.

None of these theories for the origin of the layering accounts for the erosion that seems to have occurred since the layers were deposited. There is no wind or flowing water on Io to transport material away from the retreating escarpments, and the lack of any obvious mechanism is one of the more unsatisfactory aspects in our understanding of Io's geology. A process that could explain the erosional modification is that the layered plateau deposits act as permeable channels (aquifers, as we would say on Earth) for liquid sulphur dioxide that escapes as a vapour along the escarpments. Bright deposits extending from the foot of some escarpments lend some credence to this view, if they are interpreted as sulphur dioxide frost, but the nature and origin of the impermeable cap on top of each layer that this explanation requires is unknown. This process could undermine the escarpments by sapping, but would be unable to transport the erosional debris away unless it could be blown away by high-speed gas or unless the plateau material could itself be volatilized.

Mountains occupy less than 2 per cent of Io's surface. They typically have irregular outlines and rugged, possibly fault- or fracture-controlled, surfaces, and occur at all sizes up to 200 km in diameter and 10 km in height. The mountains are generally paler and less red than the rest of Io's surface (except for the bright halo deposits) and have the appearance of protruding up through the surrounding plains unit. Haemus Mons in Plate 2 is a good example, which appears to be surrounded by a discoloured zone that may result from sulphur dioxide being channelled upwards along the boundary between the mountain-forming material and the more permeable plains material. The steepness and height of these massifs are both greater than could be sustained by a sulphur-dominated substrate, and it seems clear that the mountains, at least, represent exposures of Io's silicate lithosphere, whatever the nature of the plains and the associated lava flows.

7.1.2 Eruption plumes and hot-spots

Before delving into the debate about the surface volcanism on Io, it is worth taking a look at what can be learned from the eruption plumes. These appear to be of two main types, which are compared in Fig. 7.8. Those of the first type come from short-duration eruptions that produce plumes reaching 300 km in height and form dark annular surface deposits 1000–1500 km in diameter. The only example of this kind actually observed in progress was the plume from a source named Pele that was active during the *Voyager-1* encounter (Plates 3 and 4) but had ceased when *Voyager-2* flew by, four months later. However, new deposits that were formed around Aten Patera and Surt between the two *Voyager* encounters suggest that similar plumes had been active at these two sites in the interim. The second main type of plume, typified by that at Prometheus, is longer lived but less spectacular (Fig. 7.9). These reach heights of only 50–120 km and produce pale ring-like deposits typically 200–300 km in diameter. At least seven of the eight plumes that were active during both *Voyager* encounters are of the Prometheus type. The fissure known as Loki (Plate 5) appears to be the source of plumes of an intermediate kind; it had a plume of Prometheus dimensions at either end during both *Voyager* encounters, but during *Voyager-2*'s approach at least one of these increased to about 330 km in height for a few days before declining to about 150 km.

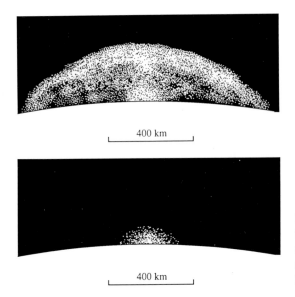

400 km

400 km

Fig. 7.8. Schematic views to compare the two types of eruption plume on Io, seen rising above the limb at the same scale. *Top*: Pele-type. *Bottom*: Prometheus-type.

The hot-spot temperatures associated with Prometheus-type plumes are less than 400 K, whereas Pele seems to have been in the 600–700 K range. The hot-spots are 20–200 km across, so the hot material is clearly localized within the source calderas. The material in the plumes themselves must chill rapidly as it rises, and has no hot thermal signature. Retrospective interpretation of ground-based infra-red observations has confirmed that Io has been volcanically active since at least the early 1970s, when the first suitable data were acquired, with a relatively steady background heat flux and short-lived bursts of energy that may be related to Pele-type events. Indeed, one such burst observed in June 1979 was probably the eruption of Surt that occurred between the *Voyager* encounters. The strongest single source of thermal radiation from Io appears generally to be Loki, which seems to have remained active up to the present time, although dramatic variations in the flux from it have been recorded. An impressive example of modern ground-based infra-red imaging techniques is shown in Fig. 7.10.

Plumes of the Prometheus type are denser than the plume observed at Pele. The plumes themselves appear dark against Io's surface, but, as we have seen, form pale deposits. The plume shape and height, and the annular nature of the deposits, are consistent with material travelling on ballistic trajectories with exit velocities at the vent of about 0.5 km s^{-1}, probably from multiple vents concentrated in a source area 20 km in diameter. In contrast, Pele-type plumes must have exit velocities of about 1 km s^{-1} to account for the height reached by the eruption. The Pele plume differed in form from Prometheus-type plumes by being umbrella-like in appearance, with the outer edge of the plume denser than the core (Fig. 7.8).

The differences between the two categories of plume in longevity, hot-spot temperature, plume dynamics, and the size and colour of their deposits suggests that there may be

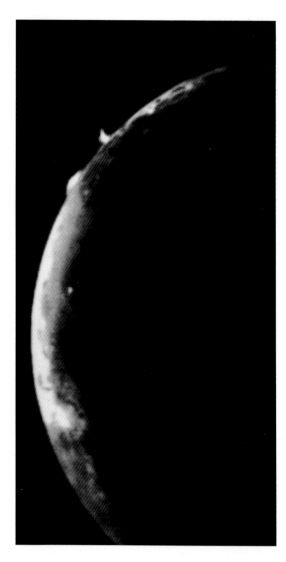

Fig. 7.9. *Voyager-2* image, looking back to Io after the probe's closest approach to Jupiter, showing Prometheus-type plumes rising above the limb from sites known as Amirani (to the north) and Maui (to the south).

fundamental differences in their origins. Volcanic eruption plumes on Earth are driven by the explosive escape of volcanic gases (dominated by water vapour) when magma comes close to the surface. In the case of Io, the apparent ubiquity of sulphur and sulphur dioxide on its surface and surrounding it in space indicates that these, rather than water, are the most important volatile phases. It has been proposed that the long lived Prometheus-type plumes are driven by the escape of sulphur dioxide heated to boiling point by the intrusion of molten sulphur (>393 K). Liquid sulphur dioxide, which exists between 198 and 263 K when subjected to a pressure of 1 atmosphere, has a very low viscosity (about half that of water) and so could move freely through the substrate, being drawn in from a wide area

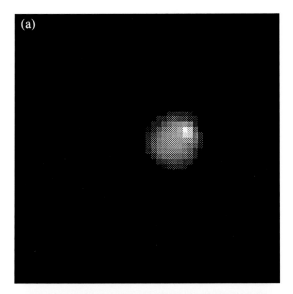

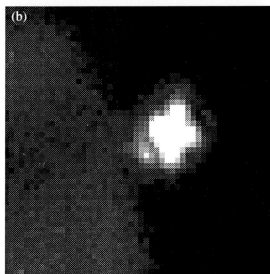

Fig. 7.10. Two images of Io on 24 December 1989 from an infra-red camera operating at 3.8 μm wavelength on the NASA Infra-red Telescope Facility in Hawaii. (a) Io in sunlight, showing excess radiation from a source at or near Loki, appearing as a bright point towards the upper right of the disc. (b) A longer exposure image recorded after Io has passed into the shadow of Jupiter. The limb of the planet can be seen, about to pass in front of Io, at the lower left. Loki is overexposed, but, because the background temperature has dropped, a second, fainter, volcano can be discerned below and to the left of Loki. Its position does not match any of the hot-spots observed by *Voyager*, so it is assumed to be a newly active volcano and has been provisionally named Kanehekili, after the Hawaiian god of thunder.

and over a long time period to replenish continuously the supply fed into a plume. In contrast, the short lived and higher temperature Pele-type plumes are more likely to be driven by vaporized sulphur. This is likely to be a result of the intrusion of molten silicate magma. The sulphur must have been heated to above 600 K, otherwise the vapour would not be able to fragment the overlying liquid sulphur, and the only material erupted would have been a frothy, vesicular flow of liquid sulphur. As liquid sulphur has a high viscosity it could not move freely through any substrate, so a Pele-type plume would be likely to deplete the reservoir of sulphur within a matter of weeks, thus explaining its short life time.

On these models the material condensing in the plumes would be dominated by sulphur dioxide in Prometheus-type plumes, which is consistent with the characteristic pale deposits, whereas Pele-type plumes would contain a lot of sulphur, which is consistent with the dark, reddish nature of these deposits.

7.1.3 Silicates or sulphur?

Thus there is evidence for the involvement of both sulphur and sulphur dioxide as major volatile phases in eruption plumes on Io. But what about the lava flows? Do these form as conventional, Earth-like flows of molten silicate rock, or do they represent surface flows of liquid sulphur? The generally red colour of Io provides no convincing evidence either way, because even if the colour is due to sulphur it could be just a thin surface coating, unrepresentative of the underlying material. Another interpretation of Io's colour that does not require large amounts of elemental sulphur is that it is due to a mixture of condensed sulphur dioxide, polysulphur oxide, and sulphur monoxide; the latter two forming as a result of the dissociation of sulphur dioxide molecules in the high-radiation environment at Io's surface.

The properties of liquid sulphur are rather exceptional, and deserve some discussion before we can attempt to assess the evidence in favour of molten sulphur at Io's surface. Under surface conditions, with no confining pressure, sulphur remains molten down to 393 K (120 °C). This is nearly 1000 K below the liquidus temperature of a molten silicate of granitic composition and 1100 K below the liquidus temperature of a silicate melt of basaltic composition. Thus, given a sufficient concentration of sulphur in Io's crust, it requires much less heat to produce a sulphur melt than a silicate melt. In silicate melts the viscosity is controlled largely by *composition* (silica-rich melts are typically a thousand times more viscous than silica-poor, basaltic melts). As a silicate lava flow moves its gross composition does not change. This means that its viscosity remains fairly constant, although it will be affected slightly by decreasing temperature, the increasing proportion of crystals present, the escape of volatiles and the confining strength of any chilled crust. In consequence, morphological differences between silicate lava flows (which may be very great) reflect their composition, and any changes along the length of an individual flow (which are more subtle) are controlled by factors such as the rate of cooling and the slope of the ground.

In the case of sulphur flows, there is no analogy with the granite–basalt silicate series or the ammonia–water mixtures discussed for icy volcanism, and sulphur flows are likely to be essentially pure sulphur. With no compositional variations to affect viscosity, it might be expected that all sulphur flows would look much the same. In fact, this is not the case because, as Fig. 7.11 shows, the viscosity of sulphur depends very strongly on temperature. Instead of increasing slightly as the temperature decreases (which is the case for most liquids) sulphur undergoes a drop in viscosity of a thousand-fold as the temperature falls from 440 to 430 K. The effect of this is that hot molten sulphur is rather viscous, similar to treacle or molasses, but as it cools it suddenly becomes much more fluid, comparable to hot engine oil. It has been suggested that the morphology of some of the flows on Io

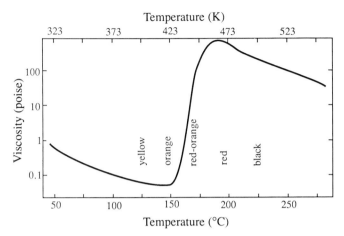

Fig. 7.11. The viscosity of molten sulphur as a function of temperature, showing also the colours of liquid sulphur in each temperature range. For comparison, the viscosity of water is about 0.01 poise and the viscosity of molten basalt is about 1000 poise.

demonstrates just such a decrease in viscosity, notably around Ra Patera (Fig. 7.12), where flows begin as very dark features around 10 km in width. These extend up to more than a hundred kilometres from the crater, becoming paler and redder, to a point where some of them suddenly fan out into broad, ill-defined orange pond-like features. The simple interpretation is that the orange ponds formed where the flow temperature had dropped below the viscosity threshold at 440–430 K, so that the suddenly decreased viscosity permitted the flow to spread out much more thinly, over a wide area. Colour changes along the flows are compatible with the changes in colour of molten sulphur, but even allowing for rapid quenching there is no agreement over how these colours could survive for any length of time under Io's surface conditions, where the stable form of sulphur would be a yellow variety.

Although some of the Ra Patera flows show moderately good morphologic evidence that they are sulphur flows, this is lacking in many other places. The dark material forming

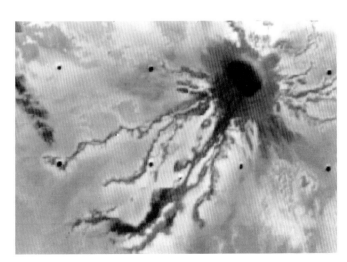

Fig. 7.12. *Voyager-1* image of Ra Patera, which is characterized by flows the morphology and colour of which change along their length (see also Plate 5). The termination of the longest flows, which reach almost to the lower edge of this view, is marked by a broad orange pond-like feature, that may represent molten sulphur becoming dramatically less viscous as its temperature fell below 440 K. The image is about 400 km across.

the floors of many of Io's calderas can be interpreted equally well as the remains of a lake of either silicate lava or of sulphur. An obvious solution to the problem is to measure the temperatures of active sites. Unfortunately, the derivation of temperatures of hot-spots at active calderas, whether using *Voyager* IRIS data or infra-red observations from Earth, is complicated by the low resolution of the detectors. Furthermore, within any volcano there are certain to be surfaces at a variety of temperatures, each making a contribution to the total infra-red flux. To take the simplest case of a lake of molten basalt, such as have been documented in Hawaii and elsewhere on Earth (Fig. 7.13), temperatures will be around 1500 K where molten lava is exposed, but most of the lake is usually covered by a chilled crust many hundreds of degrees cooler. In addition to these two volcanic components, a low resolution sensor will also detect radiation from the area surrounding the lake, which will be at the normal environmental temperature, that is about 300 K for the Earth and about 130 K for Io. The only way to untangle the signal from these three sources is to measure the infra-red flux at several wavelengths, and fit models to the resulting curve. This is no simple matter, especially considering that in reality there are likely to be gradations between these three surface temperature components, and that we can be sure of neither the temperature nor the size of any of them.

Most models of hot-spots on Io derive temperatures in the range 200–400 K, particularly for sites associated with Prometheus-type plumes, or 600–900 K for shorter lived events. The latter are taken as evidence of silicate volcanism, because some are hotter than the boiling point of sulphur (715 K at 1 bar pressure, but lower than this on Io) and all exceed the temperature that a sulphur lake boiling into a near-vacuum is likely to sustain. The data are still too crude to confirm the presence of a (presumably tiny) fraction of exposed molten silicate (>1400 K) but are compatible with a true lava lake covered with a chilled crust and surrounded by fumarolically heated ground.

Thus, although the evidence is still equivocal, it is likely that there is active silicate volcanism on Io. Certainly the steepness of the mountains (some of which have lava flows and small volcanic vents associated with them) and of the inner crater walls of the deeper calderas calls for a stronger material than could be formed by a succession of sulphur

Fig. 7.13. Oblique view into the Kupianaha lava lake on Kilauea, Hawaii (March 1990). The lake was quiescent at this time, with incandescent molten material revealed only in a crescent-shaped crack at the far side of the lake. The rest of the lake's surface was covered by a chilled crust, that occasionally broke apart to reveal more of the incandescent material beneath.

flows and sulphur dioxide frosts. On the other hand, the scalloped edges of the layers in some of the plains regions are hard enough to explain using a mixture of sulphur and sulphur dioxide, and seem very unlikely to occur in deposits dominated by silicates, so these parts of the surface are probably dominated by sulphur-rich material.

The vaporized sulphur driving the higher energy eruption plumes is likely to result from heat supplied by molten silicates, and the surface manifestations of molten sulphur in the cooler hot-spots and possibly in Ra Patera-type flows can be attributed to sulphur that has been melted and mobilized by the heat released by bodies of silicate melt cooling within the crust. On Earth, water vapour is the main means of transporting heat away from bodies of magma, so terrestrial analogues of Io-like sulphur flows and lakes of molten sulphur are rare. However, they do occur in exceptional circumstances (Fig. 7.14). It is not yet clear whether we stand to learn more about Io by studying the dynamics of such systems, or about the terrestrial examples themselves by studying Io.

Fig. 7.14. A possible analogue of sulphur lakes at Io's low temperature hot-spots: ponds of molten sulphur (partly obscured by condensing water vapour) and fresh sulphur deposits around them within the crater of Poas volcano, Costa Rica (March 1989). This view shows an area about 40 m across.

7.1.4 *Internal structure of Io*

In attempting to predict Io's internal structure a variety of factors have to be accommodated, notably: measurements of Io's size and mass, observations of its active volcanism, models of Io's past thermal and differentation history, and calculations of the present rate of tidal heating. Io's density (Table 1.1) shows that it is essentially a rocky body. Its present waterless nature can be explained by starting with a bulk composition equivalent to carbonaceous chondrite meteorites but excluding volatiles such as methane, carbon dioxide, and water, due to the relatively high temperature environment in which Io formed, and allowing a combination of radiogenic and tidal heating to produce temperatures high enough for convection and outgassing (and ultimate loss) of any volatiles that survived accretion. A silicate lithosphere would have formed, and flexing between this shell and the underlying asthenosphere would be the main means of heating by the dissipation of tidal energy.

Early models that suggested that the interior of Io is entirely molten have now been discounted. It is considered more likely that Io must have differentiated to produce an iron–sulphur core overlain by a silicate mantle. The lower mantle is probably solid and relatively rigid, but there must be a convecting asthenosphere, and possibly even a thin liquid shell, below the lithosphere. Sulphur, of which there is about 5 per cent by mass on the dehydrated chondritic model, would be sufficient to give a 50 km thick crust. Clearly this is far more than is required to explain the nature of the surface, and some would probably remain trapped in the core. A reasonable model for the internal structure of Io, including some differentiation of the lithosophere, is illustrated in Fig. 7.15.

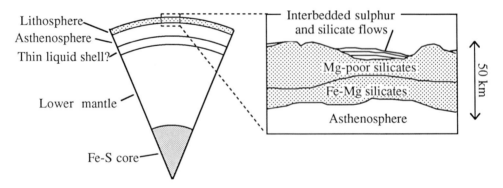

Fig. 7.15. Model of the internal structure of Io. The core may be molten. In the view on the left, the thicknesses of the layers above the lower mantle have been exaggerated for clarity. The detail of the lithosphere on the right is drawn approximately to scale, although its thickness is uncertain by a factor of at least two. The iron–magnesium silicate layer is analogous to the lithospheric part of the Earth's mantle, and the magnesium-poor silicates and overlying flows constitute the crust. Intrusions of silicate melt into the crust probably power the sulphur- and sulphur dioxide-driven plumes.

Evidence that differentiation within Io's lithosphere has produced a distinct crust is given by the rapid rate at which Io is being resurfaced. The lack of impact craters requires that new material is added to the surface at a rate in excess of 0.1 cm yr^{-1}. The heat loss from hot-spots, averaged over the globe, is equivalent to resurfacing at a rate of about 1 cm yr^{-1} if all the flows are silicate and about 10 cm yr^{-1} if all the flows are sulphur, to which plumes would add up to 1 cm yr^{-1}. At rates such as these the whole mass of Io could have been recycled during its history, or, more probably, the lithosphere and upper mantle could have been recycled many times. This vigorous activity is not accompanied by any sign of global faulting and deformation comparable with plate tectonic processes on Earth; such faults as are present are attributable to tidal flexing. Presumably, if Io is in a steady state, then any fresh silicates added to the crust by volcanism resulting from the rise of melts from the asthenosphere must be compensated for by the base of the lithosphere being eroded away at an equivalent rate by reincorporation into the asthenosphere.

Despite the lack of plate tectonic processes, there are indications that neither crustal compositon nor the distribution of activity on Io are uniform. Most of the large mountains on Io that have been mapped are between longitudes 230–360°. This may be an artefact, because this region coincides with the highest resolution images, but if it is real and the mountains represent protrusions of silicate crust through a mixed sulphur and silicate layer, as implied in Fig. 7.15, then the crust is likely to be thicker in this region by several kilometres. Interestingly, Pele-type plumes and most of the hot-spot heat flow observed since the *Voyager* missions are concentrated in the same general region. The total heat flow from Io is estimated at about 1–3 W m^{-2} as a global average, compared to a global average of 0.06 W m^{-2} for the Earth. This is in fact more than can be easily accounted for by current models for Io's tidal heating. The present plume and heat-flow patterns may not persist in the long term, and it is possible that their distributions could change shortly; however, observations show that what we now know to be the more active hemisphere has been redder than the other for at least the past 60 years. There is no alternative but to continue ground-based monitoring such as that illustrated in Fig. 7.10, and to await further high-resolution imaging of Io by spacecraft. This would show whether the apparent asymmetry in the distribution of mountains is real, and would also reveal changes in plume and flow deposits.

Much still remains to be understood about Io, and we are fortunate that certain things such as the cloud of neutral and ionized gases surrounding it, its spectrum and hence the composition of its surface, and the size, temperature, and location of the major hot-spots can be measured and monitored from Earth. When we turn to Io's neighbour, Europa, we find a world that is almost equally as intriguing but which veils its mysteries in ice, giving us much less to work on.

7.2 Europa

With Europa, the smallest of the galilean satellites, we make a partial return to the icy worlds that predominate in the outer solar system. Its density is marginally less than that of our own Moon (Table 1.1), but it is not at all Moon-like to look at (Fig. 7.16). It has almost no craters and its appearance is dominated by dark crack-like features. Both attributes are clear signs of very young, and probably continuing, geological activity.

Europa's albedo is high and its spectrum is dominated by fairly clean ice (Fig. 2.1) demonstrating less contamination by silicates than Ganymede and Callisto. To account for Europa's density there must be at least 5 per cent of water mixed in with silicate rock. If all the water were at the surface it would form a layer over 100 km thick, and somewhat

Fig. 7.16. *Voyager-2* view of part of Europa. Mottled terrain and plains areas are seen cut by a variety of mostly dark narrow features. Two craters are barely visible in this image; a bright 15 km diameter crater with a dark halo (A) and, in the lower right-hand corner, a larger multi-ringed crater (B). Tyre Macula, the dark spot (C) in the upper right, may be a crater palimpsest.

less than this if the upper part of the silicate mantle were hydrated. Considerations of radiogenic and tidal heating, together with the lack of impact craters and the general low relief of the surface, which we shall discuss shortly, show that the lower part of the water layer is likely to be liquid. It has even been suggested that geothermal and tidal heat from below and sunlight penetrating the ice from above make this global ocean on Europa the most likely abode of life in the solar system beyond the Earth. A reasonable model of the compositional layering within Europa is shown in Fig. 7.17.

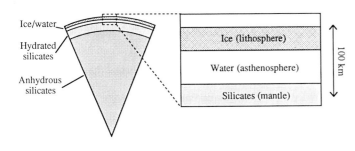

Fig. 7.17. Model for the internal structure of Europa. The extent of hydration of the silicates is uncertain, as are the relative thicknesses of the ice and liquid water layers. However, the icy lithosphere is likely to be about 10–30 km thick. Other models have ice resting directly on silicates, with no intervening water layer; even so, the lower part of the ice is likely to be warm and to have a viscosity sufficiently low for it to act as an asthenosphere.

7.2.1 The surface of Europa

Europa was the most poorly imaged of the galilean satellites (Fig. 3.6). The resolution of the best images is tantalizingly just too coarse to be certain of many of the details, although the data are adequate to establish the general nature of Europa's surface. There are two distinct terrain types on Europa: a bright plains unit and a darker (mostly brown) mottled terrain. Both these units are cut by narrow linear and curved bands that are generally dark, although some have a bright stripe along their centre (several examples can be seen in Fig. 7.16) and there are a few that are almost entirely bright (Fig. 7.18). Many of the bands

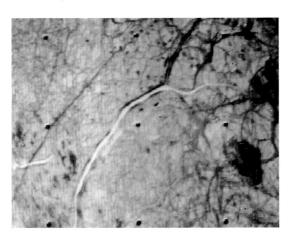

Fig. 7.18. *Voyager-2* image covering Agenor Linea, a bright band on Europa. The area shown is about 1000 km across. The dark patch in the lower right appears again in Fig. 7.19.

can be traced for several hundred kilometres, although there are shorter, mostly younger, wedge-shaped examples that we shall turn to shortly. Superimposed on these bands and terrain types are long ridges only a few kilometres wide and about 200 m high that are the youngest class of tectonic feature on Europa. These are best seen near the terminator where the low sun angle accentuates them. In the southern hemisphere in particular, each ridge typically has the form of a succession of arcs joined together (Fig. 7.19).

These cycloid ridges are unlike anything else known in the solar system, and Europa is also remarkable in its extraordinary low relief. In the mottled terrain there appear to be irregular depressions with steep walls but without raised rims, and some low domes. Some of the dark and bright bands may be subtle ridges or troughs, although none of them exceed a few hundred metres in relief. Craters are very rare; less than a dozen between about 5 and 30 km in diameter have been identified with any certainty. A few roughly circular brown spots 100 km in diameter, such as Tyre Macula (Fig. 7.16), may be the scars left by relaxed craters (like the palimpsests on Ganymede), except that they have absolutely no discernible relief.

7.2.2 Geological history

There is no direct evidence of geological activity on Europa, except for a low resolution (40 km per pixel) *Voyager-2* image that shows what could be interpreted as an active water eruption plume rising above the crescent limb in one corner of the frame. This interpretation has been hotly disputed, and the consensus is that the supposed plume is just noise. Nevertheless, Europa's surface is clearly one of the youngest in the solar system. The density of craters with diameters of more than 10 km on Europa is way below that seen on the lunar maria, and yet gravitational focusing by Jupiter makes it likely that Europa has experienced a greater flux of impacting bodies than the Moon over recent times. Building on this logic, the best estimate is that the average age of Europa's surface is no greater than ten to a hundred million years. Craters must, therefore, be being removed by resurfacing (with help from viscous relaxation) at a considerable rate, and it is logical to suppose that the processes responsible continue today. The conundrum of which of these processes is the more important is unlikely to be solved until we have higher resolution images; a disproportionately large number of small craters (around 1 km in diameter) would indicate that the larger craters had been at least partly removed by viscous relaxation, thereby providing a clue as to the thickness of the lithosphere. In contrast, if a size–frequency distribution plot showed the smaller craters to fall on the same trend as the larger craters, this would indicate that the overall dearth of craters is due to resurfacing.

The areas of mottled terrain appear to be superimposed on the plains units, so the plains must be the oldest terrain preserved on Europa. These plains are evidently surfaced by relatively clean ice. They could represent a crust frozen out of a global water ocean at the end of a melting event (perhaps associated with a change in orbital tidal interactions between the inner galilean satellites, which could have happened a few hundred million years ago). Another possibility is that the plains are underlain by old, viscously relaxed crust that has been mantled by widespread water or ice flows, or pyroclastic frost deposits

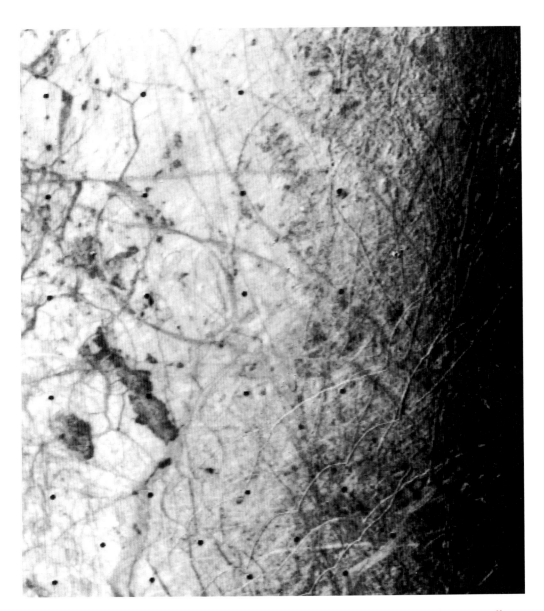

Fig. 7.19. *Voyager-2* view of part of Europa's southern hemisphere. Cycloid ridges are well seen in the lower right, and a variety of irregular pits and domes can be made out in the mottled terrain towards the upper right. Dark wedge-shaped bands occur in the upper left.

from explosive eruptions. The mottled terrain, and various smaller irregular brown patches, could represent areas where warm ice contaminated by silicate dust or other debris forms shallow intrusions or has been extruded across the surface. In these regions the ice layer may rest directly on the underlying silicates, with no intervening layer of water. The presence of contaminated ice below the surface is supported by the brown mantle of ejecta surrounding crater A in Fig. 7.16.

Europa's various narrow belt-like features began to be formed at the time the mottled terrain was developing, and they continued to be produced afterwards. Mutual cross-cutting relationships demonstrate that the triple bands are generally the oldest. One interpretation is that these are fractures filled with a brecciated mixture of ice and rock, the central bright stripe having frozen from relatively pure, frothy water that was extruded last. An alternative model regards the bright stripe as the primary crack-filling feature with the dark zones on either side representing some kind of associated surface deposit. Similarly, the dark bands are usually regarded as crack-filling features. The bands in general are clear evidence of resurfacing at least in local areas.

In one region of the southern hemisphere (extending into the lower left of Fig. 7.16) the dark bands are cut by a family of shorter, wedge-shaped bands that demonstrate an even younger episode of activity and are examined in the next section. Possibly the youngest event of all was the creation of the cycloid ridges. These overprint all the pre-existing terrain features, but they do not occur in the same region as the wedge-shaped bands and so the relative ages of these two phenomena remain unknown. Attempts to interpret the surface features of Europa are dogged by the restricted global coverage of the images, and their inadequate resolution. This means, for example, that we cannot be sure what happens when one feature cuts across another, nor can we resolve the internal structure of even the broadest triple bands. It is clear that Europa has had a complex history, of which we have as yet a very poor understanding. To try to get some feeling for the kinds of processes that have occurred, we will conclude our look at Europa with a brief examination of its tectonics, paying particular attention to the wedge-shaped bands.

7.2.3 Tectonics on Europa

With the exception of the cycloid ridges, the story we can see of Europa's tectonism, at least in the 25 per cent imaged at 2 km per pixel, is one dominated by extension. Whether they are grabens or simply cracks, the dark bands in particular are hard to interpret convincingly except as extensional features. Global expansion is one possible cause, which could be due to dehydration of Europa's centre to form an anhydrous core leading to increased hydration and expansion of the overlying silicate mantle. On the other hand, there have been various attempts to relate the pattern of fractures on Europa to stresses caused by tidal forces, more recently taking account of the possible decoupling of the ice lithosphere from the interior to allow it to rotate non-synchronously despite the seemingly inevitable tidal locking of the silicate interior.

An alternative model is that Europa's tectonic activity is driven by convection in the ice–water asthenosphere, in a manner comparable to (although probably not identical to) plate tectonics on the Earth. The major bands whose length approaches 1000 km and which lie along great circles are probably best interpreted as due to a global cause (such as expansion or tidal forces), but the wedge-shaped bands are an attractive example of the scale of feature that would be likely to result from convection in an asthenosphere of the thickness shown in Fig. 7.17, irrespective of whether this asthenosphere is warm ice or liquid water.

One of the more intriguing parts of the wedge-shaped band province is shown in close-up in Fig. 7.20. Here there are several interconnected wedge-shaped bands with convincing evidence that they cause offsets on pre-existing bands. An interpretation of this region is given in Fig. 7.21. Here there is evidence that the icy lithosphere is broken into

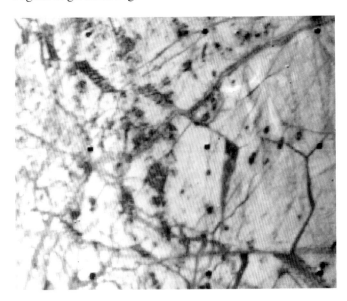

Fig. 7.20. Detail of part of a *Voyager-2* image from an area overlapping the top left of Fig. 7.19. Wedge-shaped bands cause offsets on pre-existing dark bands and a triple band. The tectonics of this region are interpreted in Fig. 7.21.

Fig. 7.21. Tectonic interpretation of the area shown in Fig. 7.20. The sketch has been simplified by omitting some superimposed brown patches and older fractures. Wedge-shaped bands are interpreted as marking zones of extension. The linked segments of wedge-shaped bands extending from x to y are offset by the equivalents of transform faults. A pre-existing triple band and three very narrow dark bands have been displaced in a right-lateral sense across the first wedge-shaped band extending north-westwards from y, due largely to spreading by the wedge-shaped band extending south from y. There has been a relative rotation between plates A and B of about 5°, as implied by the southwards narrowing of this feature. Note that if plate B were rotated clockwise to close up the gaps represented by the wedge-shaped bands on its northern and eastern boundaries, the displacements suffered by the narrow dark bands and the triple band would be removed. In addition, plates A, B, C, and D have each migrated progressively further to the west (relative to the north-east part of the area), as implied by the lengths of the arrows.

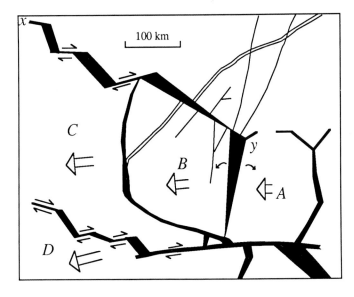

plates of the order of 200 km across, the relative motions between which are shown by the shape and configuration of the wedge-shaped bands separating them. Those bands that are truly wedge-shaped provide evidence of relative rotation between plates, and the amount of this rotation is in agreement with the amount of displacement experienced by the older bands that they cut. Furthermore, segments of wedge-shaped band can be seen to be offset by narrow features that appear analogous to oceanic transform faults on Earth.

It should be emphasized that even if this interpretation is correct, Europa does *not* show Earth-style plate tectonics. However, with the possible exception of Venus, it appears to have the nearest thing to it that we are aware of on any other planetary body. The analogy between plate *B* on Fig. 7.21 and the Easter Island microplate on Earth (Fig. 7.22) is striking, although there is an important difference: on Europa the total amount of spreading (represented by the wedge-shaped bands) is small, whereas once spreading has started in the Earth's oceanic lithosphere it is liable to continue indefinitely.

By summing the widths of Europa's wedge-shaped bands across the wedge-shaped band province, the total amount of extension can be estimated at nearly 100 km. Unless global expansion is allowed, there ought to be evidence of the destruction of an equivalent amount of lithosphere elsewhere. The bright band Agenor Linea (Fig. 7.18), which also appears to be a young feature, has been suggested as a possible candidate. It does not look like the trace of a terrestrial subduction zone, but it may be a pressure ridge of some kind. The cycloid ridges offer another possibility, and parallels have been drawn between them and the strings of island arcs in the western Pacific Ocean that overlie subduction zones, where the destruction of oceanic lithosphere compensates for spreading in other parts of the ocean. On Europa, however, there is no obvious geometric relationship between the directions of spreading implied by the wedge-shaped band province as a whole and the trend of the cycloid ridges.

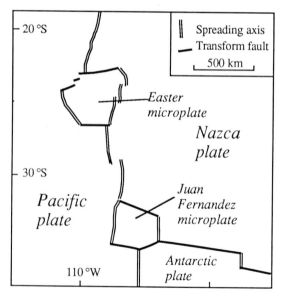

Fig. 7.22. Plate tectonic map showing the spreading axes and transform faults bounding the Easter microplate in the Pacific Ocean, of which plate *B* in Fig. 7.21 is a possible europan analogue.

A resolution of the mysteries of tectonics on Europa and the question of at what rate tectonic and volcanic processes continue there today must await more detailed and more widespread imaging by future missions. By revealing any smaller craters that there may be, images of higher resolution ought to cast considerable light on the relative importances of resurfacing and relaxation of topography. In addition, more images taken at low Sun angles would be a great help in unravelling the topographic expressions of the host of narrow bands that cut the face of this most intriguing world.

7.3 Enceladus

The grooved terrain on Ganymede is unique in the solar system. The closest we find to it elsewhere is on Enceladus (Fig. 7.23, Plate 6), where there is an old cratered terrain cut by and overlain by smoother units that have concentrations of grooves. This, however, is where the analogy ends, because in other ways Enceladus is a very different world to Ganymede. For example, Enceladus is at the opposite end of the size and mass range (Table 1.1), and in that respect is much closer to Mimas; but Mimas, as we saw in Chapter 5, is a heavily cratered little world showing no signs of endogenic activity.

7.3.1 Terrain units on Enceladus

Although we have useful images of less than half its surface, Enceladus displays the widest variation in crater densities and morphology of all the satellites of Saturn. Craters of comparable size may vary from fresh bowl shapes to highly degraded forms with low rims and bowed-up floors, as can be seen in Fig. 7.23. This is quite unlike Mimas, where there is no viscous relaxation. Six distinct terrain units have been identified on Enceladus, and these are distinguished on Fig. 7.24. There are three cratered units, which, in order of decreasing age are: cratered terrain 1 (ct_1), having highly flattened craters in the 10–20 km size range; cratered terrain 2 (ct_2), having well preserved craters in the same size range; and the cratered plains (cp), which are characterized by a lower crater density and bowl-shaped craters in the 5–10 km size range. The next youngest unit is smooth plains 1 (sp_1), which is cut by linear grooves and has a light sprinkling of craters. This is succeeded by smooth plains 2 (sp_2), which is similar to sp_1 but has no detectable craters. Around the margins of sp_2 an uncratered ridged plains unit (rp) has developed, which can be seen to have overwhelmed ct_2 along a sharp boundary where pre-existing craters are truncated.

The density of craters in ct_1, the most densely cratered unit on Enceladus, is about the same as that on the least-cratered areas of Saturn's other satellites. This implies an age of probably no more than about 3.8 billion years, and considerably younger than this if the likelihood of Enceladus having been broken apart by impacts and then re-accreted is to be taken seriously (see section 5.6). Whatever the merits of that argument, the absence of craters on the sp_2 unit down to the observable limit on the best images we have (about 2 km) means that these are very unlikely to be more than a few hundred million years old.

The albedo of Enceladus is nearly uniform across all terrain units, and at virtually 1 is the highest known in the solar system. Analysis of the way the surface scatters light suggests that it is covered by fresh frost. A likely cause of this is globally dispersed spray from volcanic eruptions, perhaps associated with the generation of the sp$_2$ terrain. As the surface coating is uncontaminated by a detectable amount of meteoritic dust or other rocky particles, it must be very young and it is likely that some form of geological activity is continuing on Enceladus up to the present day. The most compelling argument in support of this is that Saturn has a very diffuse outer ring, known as the E ring, that extends between 210 000 and 300 000 km from the centre of the planet. Photometric studies suggest that this ring consists of spherical particles about a micrometre in size, and that

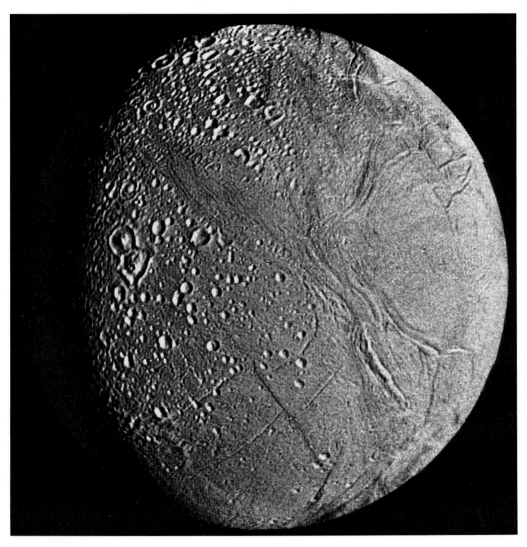

Fig. 7.23. High resolution mosaic of *Voyager-2* images of Enceladus. A colour version is shown in Plate 6.

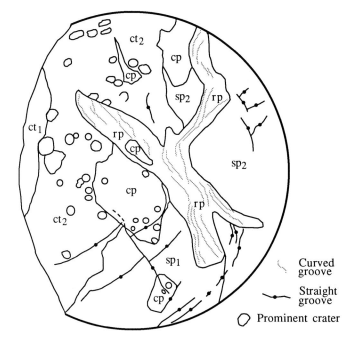

Curved groove

Straight groove

◯ Prominent crater

Fig. 7.24. Geological sketch map of terrains on Enceladus (same area as Fig. 7.23). See text for discussion.

these particles reach a maximum concentration at about 230 000 km. As Table 1.1 shows, this distance coincides with the orbit of Enceladus. This suggests that Enceladus is the origin of the E ring particles, and it is tempting to attribute them to spray from explosive volcanic eruptions, just like the surface frost. It has been argued that these particles would not survive in such a ring for more than about ten thousand years. If this is correct then the case for continuing activity of some kind on Enceladus is very strong.

7.3.2 Volcanic and tectonic processes on Enceladus

Whatever the merits of arguments for present-day activity on Enceladus, it has clearly had a prolonged active geological history, with several episodes of resurfacing. The variable degree of degradation of crater morphology within terrain units suggests that as well as changing over time, the heat flow may have varied from place to place. Perhaps the local areas of extreme crater degradation once lay above sites of convective upwelling in the asthenosphere.

There are essentially two types of grooves on Enceladus, both of which can be made out on Fig. 7.23. The first type consists of single straight grooves that lie within both the cratered terrain and plains units. These are up to about 100 km long, and typically 2–4 km across and a few hundred metres deep. The other type of groove tends to occur in sub-parallel swarms that are curvilinear and are concentrated in the ridged plains unit, which lies along the edge of the sp_2 unit and in a wedge bounded by cratered units. Curvilinear groove sets can be traced for up to 200 km.

The straight grooves are generally interpreted as tensional features. Their narrowness and the low gravity of Enceladus make it likely that they are simple extensional fractures, rather than grabens consisting of down-dropped blocks between two faults. Enceladus is too small for ice II to exist in the interior, and so the likeliest cause of surface extension is global expansion due to the freezing of water in the interior, to form ice I, which is several per cent less dense than water. There are, however, other plausible explanations for some of these grooves. One of the most striking features on Enceladus is the 20 km offset of the straight groove known as Daryabar Fossa where it crosses the straight groove Isbanir Fossa in the sp$_1$ terrain in the lower part of the disc in Figs 7.23 and 7.24. It is tempting to make an analogy with an oceanic spreading axis on Earth (essentially a tensional fracture) offset by a transform fault, as shown schematically in Fig. 7.25a. On this model, Daryabar Fossa has acted as a spreading axis, responsible for the generation of the smooth plains

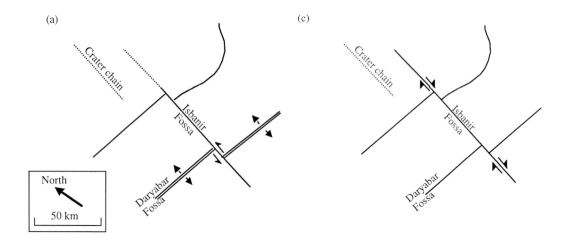

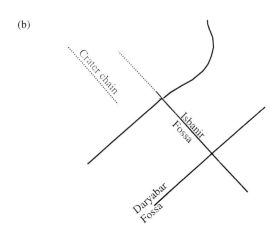

Fig. 7.25. (a) Simplified tetonic map of the area including Daryabar Fossa and Isbanir Fossa on Enceladus, indicating these two fossae, an un-named linear groove to their north, and a crater chain. The arrows show the direction of relative movement if Daryabar Fossa is interpreted as a spreading axis and Isbanir Fossa as a transform fault. Parts (b) and (c) show two stages in an alternative model in which Isbanir Fossa is a transcurrent fault that forms after Daryabar Fossa and the un-named groove, and causes the same offset on both. The relative age of the crater chain is not constrained by either of these models.

unit within which it lies, and Isbanir Fossa would be a transform fault and fracture zone, formed by sideways slip of lithospheric plates generated at the spreading axis. Alternatively Isbanir Fossa might be a conventional transcurrent fault with 20 km of right-lateral motion that caused the offset on Daryabar Fossa, which would then have to pre-date Isbanir Fossa (Fig. 7.25b and c). This explanation is favoured by the offset in the same sense and direction that Isbanir Fossa appears to cause on a second linear groove within the cratered terrain 50 km to the north, but is seemingly contradicted by the fact that Isbanir Fossa terminates abruptly in old terrain at either end, whereas a transcurrent fault should be traceable indefinitely until it reaches another structure than could compensate for the displacement.

Yet another explanation that has been put forward for the straight grooves is that they are volcanic fissures. This idea is favoured by the existence of at least one crater chain, parallel to Isbanir Fossa, which appears to have been formed by explosive eruptions concentrated at discrete locations along a linear feature, presumably a fissure that penetrated down to a molten or partially molten interior. It is not surprising to find signs of volcanism along fractures such as this: any line of weakness in the lithosphere would be a pathway up which melts could travel. As outlined in section 4.2.3, the presence of ammonia or other volatiles in either the melt or the ice of the fissure walls could stimulate an explosive, pyroclastic eruption in several ways. Although the presence of significant amounts of voltatiles in Enceladus' ice has not been independently demonstrated, such eruptions are an attractive proposition because, in view of the low surface gravity of Enceladus, they could be a source for global frost deposits and even the icy droplets in Saturn's E ring. Quieter, more effusive eruptions could have led to the formation of the plains units.

The belts of curvilinear grooves, forming the young (uncratered) ridged plains unit are another story. Each groove could represent a fissure, a narrow graben or some other volcano–tectonic feature formed by one of the processes suggested for the grooved terrain on Ganymede or the belts of ridges in the coronae of Miranda. The wedge of ridged plains unit that intervenes between areas of cratered terrain (towards the upper left in Figs 7.23 and 7.24) certainly has the appearance of an extrusional–tensional feature. However, the belts of ridged plains unit that wrap around the edge of the major areas of smooth plains (in Figs 7.23 and 7.24, and in the area out of sight in this image, beyond the upper right limb) are more problematic. They may have formed in one of the ways suggested for the straight grooves and the other ridged plains area, but it is tempting to see them as an example of compressional terrain deformed between more stable blocks. A possible analogy is with a greenstone belt in a terrestrial granite–greenstone terrain (e.g. Fig. 7.26).

7.3.3 *Heat source for Enceladus*

The very young and probably continuing activity on Enceladus is in marked contrast to the situation on Dione and Tethys, which, as we saw in the previous chapter, show clear evidence of past (but not continuing) geological activity. As neither of these larger and equally dense worlds shows the effects of radiogenic heating, it would be unreasonable to

Fig. 7.26. A tempting analogue for the belts of ridged plains wrapping round the edges of smooth plains on Enceladus (Fig. 7.23). This is a satellite image covering part of Western Australia, about 30 km across. The bright areas at lower left and lower right are undeformed granitic cratons and the dark terrain separating them is greenstone belts. This configuration is inherited from 2–3 billion years ago, when the Earth's lithosphere was considerably thinner than today. The greenstone belts are compressed volcanic terrains that may have originated as small ocean basins.

appeal to this as the cause of the activity on Enceladus. Models for the thermal evolution of Enceladus based on both radiogenic heat alone and on radiogenic heat supplemented by accretionary heat show that its history should be intermediate between those predicted for Rhea (Fig. 5.17) and Mimas (Fig. 5.20), and, in view of its small size, it ought to be more similar to Mimas than to Rhea. The remaining source of heat to which we can appeal as the powerhouse for the activity on Enceladus is tidal dissipation.

We have already seen that the dissipation of tidal energy is an extremely effective way of keeping the interiors of Io and Europa hot. A quick examination of the orbital resonances between the saturnian satellites suggests considerable potential for such a mechanism in Enceladus, because the orbital period of Enceladus is exactly half that of Dione. This means that Enceladus experiences a gravitational 'tug' from Dione twice in every orbit. Unfortunately, however, the observed eccentricity of Enceladus' orbit, which is presumably caused by this resonance, is rather small. According to models based on our (admittedly poor) understanding of the mechanical properties of volatile-rich ice under high pressure, the rate at which this would generate heat by tidal dissipation is too low to cause melting. One way out of this impasse is to appeal to changes in orbital resonances over time; after all, Saturn's satellite system is very complex, and gravitational interactions between more than two bodies are notoriously difficult to calculate.

The most likely multi-body scenario can be imagined by considering Janus and Epimetheus. These are a pair of small low density irregular-shaped satellites each about 150–200 km across, whose orbits lie inside that of Mimas (they are shown but not named on Fig. 1.2). Their orbital periods are only fractionally more than half that of Enceladus, and are apparently changing as a result of gravitational interaction with Saturn's rings. If either of these were to have been temporarily locked in a 2 : 1 orbital resonance with Enceladus, while Enceladus remained in a 2 : 1 orbital resonance with Dione, the forced eccentricity of Enceladus' orbit should have increased to an extent where melting would have become feasible. It is possible that Enceladus has experienced several periods of forced orbital eccentricity that lasted, say, between ten and a hundred million years each time, and that each of these was marked by the resurfacing of part of the globe. Thus it appears that we can explain past episodes of activity on Enceladus if we admit that we do not understand the complex orbital interactions. However, this does not account for any present day activity very well. The alternatives seem to be to regard such activity as the waning phase of eruptions caused by the most recent episode of internal melting, or to accept that there is no continuing activity and attribute the frost to something that happened millions of years ago and leave the origin of the E ring unaccounted for.

Whether or not Enceladus is geologically active now, the facts that there have been several episodes of activity with the youngest occurring in the geologically recent past (as attested to by the lack of craters on the sp$_2$ unit), and that the only effective heating mechanism we can think of is apparently episodic, make it a pretty good bet that Enceladus will 'switch on' again at some time in the geologically close future, within a few hundred million years, if not before.

If the grounds by which Enceladus wins its place in the 'active worlds' category are debatable, there can at least be no argument about the final example, Triton, which is the second moon on which clear signs of eruptive activity in progress have been seen.

7.4 Triton

With Triton we reach the most distant world visited by any space probe from Earth. It was a fitting climax to the *Voyager* project that Triton turned out to be such a fascinating place, with a geologically young and active surface (Plate 7). During one week at the end of August 1989, as *Voyager-2* skimmed 5000 km above Neptune's cloud tops and then on to within 25 000 km of Triton, the attention of the world's media was focused on planetary exploration, with an enthusiasm not seen since the *Apollo* Moon landings.

Lying on the outer fringes of the solar system, Triton is an icy world, the like of which we have not encountered before. It is large (Table 1.1), beaten in size only by Titan and the galilean satellites of Jupiter, and is remarkable in being the only major satellite of any planet to have a retrograde orbit. That is to say, its orbital motion is in the opposite direction to its planet's spin, and in this case it is inclined at 21° to the planet's equator. As discussed in Chapter 2, its retrograde orbit makes it virtually impossible for Triton to have accreted from a protosatellite disc around Neptune, so it has long been thought likely that Triton must have formed elsewhere. Evidently, it eventually wandered close to Neptune, where collision with another satellite or frictional drag in any remaining gas cloud about the planet could have slowed its motion down sufficiently to allow it to be captured into orbit. Initially its orbit would have been highly eccentric, but tidal interaction with Neptune would eventually have forced it into its present circular pattern. This process may have started immediately upon capture or it could have taken a billion years or more to begin; either way it would probably then have taken of the order of a hundred million years to run to completion. While the orbit was being changed, dissipation of tidal energy must have acted as a major heat source (possibly exceeding radiogenic heating by a factor of a thousand or more), leading to global melting and differentiation.

7.4.1 *Composition of Triton*

Triton's faintness and proximity to Neptune make it difficult to study from Earth. Prior to *Voyager-2*'s encounter, methane had been identified in Triton's spectrum, although it was not clear whether it took the form of ice or gas. There was no clear trace of the water-ice absorption features that are so prominent elsewhere, and one faint spectral feature inspired the hope that parts of Triton were covered by oceans of liquid nitrogen. These would have made a fascinating sight, but the reality when *Voyager-2* got there was almost as exciting. The surface temperature turned out to be about 38 K, rather colder than expected (indeed colder than any other observed surface in the solar system), and sufficiently low to freeze nitrogen solid. The presence of scarps up to 1 km in height demonstrates that the surface has the strength of water-ice, but it is evidently covered by a mixture of methane and nitrogen frost, giving it a generally pinkish appearance. There is a polar cap, apparently dominated by nitrogen ice, that may be a largely seasonal phenomenon. The whole of Triton's surface has a high albedo, with a global average of 0.78, which is exceeded only by that of Enceladus. Even the darkest material on Triton is really quite reflective, with an albedo around 0.62, whereas the brightest parts of the polar cap have an albedo of 0.89. Triton has a very thin atmosphere, mostly nitrogen with some methane, with a surface pressure of 16 ± 3 μbar, not much more than a hundred thousandth of the Earth's atmospheric pressure. The atmosphere contains rare thin clouds at altitudes of a few kilometres that were seen only over the polar cap and are interpreted as condensed particles of nitrogen-ice (analogous to terrestrial cirrus clouds which are of water-ice crystals). There is also a pervasive, although tenuous, haze extending to a height of 30 km that is probably a photochemical smog consisting of ethane, acetylene and the like, resulting from the action of solar ultraviolet radiation on methane.

We shall examine some details of Triton's surface shortly, but first it is worth considering its bulk composition and internal structure. The *Voyager-2* encounter showed that Triton's density exceeds that of Ganymede and Callisto (Table 1.1). Bearing in mind that its lower mass must result in less internal compression, such a high density can be explained only by Triton's having a considerably greater rock to ice ratio than these. In fact, silicates must make up between about 65 and 75 per cent of the total mass, greater than for any other major outer planet satellite except Io and Europa.

In Triton's case, it is simplest to regard this ratio as reflecting a decrease in the amount of water available to condense from the solar nebula rather than an excessive abundance of silicates, because the latter would be contrary to the solar nebula condensation models discussed in Chapter 2. The low abundance of water can be accounted for by arguments based on chemical reaction rates, which suggest that in the cold outer region of the solar nebula where Triton probably formed (and in the absence of local additional heating from the condensation of a neighbouring gas giant planet), much of the hydrogen and oxygen would have been partitioned into the phases H_2 and CO (hydrogen and carbon monoxide) in preference to H_2O and CH_4 (water and methane). With relatively little water available to condense when Triton formed, the rock to ice ratio was increased. The same situation would favour N_2 (nitrogen) over NH_3 (ammonia). The big problem with Triton's composition is that of the missing carbon monoxide. This species has not been detected on Triton, which means that its abundance at the surface must be less than about a tenth that of nitrogen.

Notwithstanding the temperature-related compositional effects, a combination of theory and observational evidence suggests that the bulk of the ice now found on Triton is water-ice. Triton is now almost certainly fully differentiated, if not as a result of its original accretional and radiogenic heat, then because of the post-capture tidal heating event. A model of its possible structure is shown in Fig. 7.27. The tidal heating event would probably have sustained the existence of a mantle of liquid water for a considerable period, although it is likely to be almost entirely frozen by now. The water mantle stage of Triton's history may contain a clue to the missing carbon monoxide: while the 'ocean' existed, tidal and radiogenic heat in the core could have been carried away convectively by the drawing down of cold water through fractures in the rock, whereupon it would have been heated to escape back into the ocean through hydrothermal vents. While this was going on, carbon monoxide that was in solution would have had the opportunity to react with the warm, hydrated rock. The sort of chemical reaction that could have then occurred can be summarized thus:

$$CO + 2H_2O + rock = CH_4 + oxidized\ rock$$

Interestingly, methane is the most abundant gas dissolved in the water escaping from hydrothermal vents such as black smokers at active spreading axes in the Earth's oceans today (Fig. 7.28). However, most of this is primordial methane that has been trapped within the Earth since its accretion and little, if any, is the immediate product of the reduction of carbon monoxide as proposed for Triton.

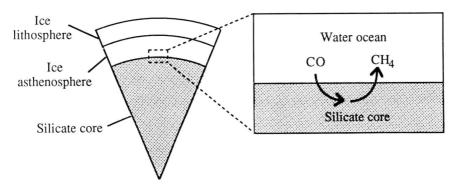

Fig. 7.27. Model for the present internal structure of Triton. Radiogenic heat from Triton's large rocky core is probably sufficient for the melting point of water-ice to lie near the core–mantle boundary, so the lithosphere–asthenosphere transition is likely to be no deeper than the middle of the mantle. The presence of ammonia and/or methane as partial melts may decrease the thickness of the lithosphere still further. The detail shown on the right indicates the situation during tidal heating, when the water-dominated mantle was largely molten and the circulation of hot aqueous fluids through the outer part of the silicate core could have lead to the chemical reduction of carbon monoxide (CO) to methane (CH$_4$).

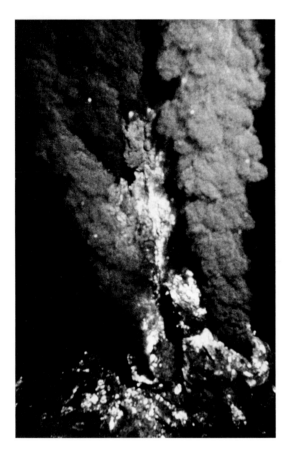

Fig. 7.28. Hydrothermal vents of the type known as black smokers, seen here on a spreading axis on the floor of the eastern Pacific Ocean. Water heated by passage through underlying hot rocks, with which it has interacted chemically, escapes at about 350 °C from vents about 20 cm in diameter. Upon mixing with the ambient sea water, a black precipitate of metal sulphide particles is created, giving rise to the opaque plumes. Comparable chemical interchanges between water and rock may once have occurred on Triton.

7.4.2 *The surface of Triton*

Owing to the high inclination of Triton's orbital plane the northern polar region was in shadow in 1989. *Voyager-2* imaged as much of Triton's Neptune-facing hemisphere as was sunlit with a best resolution as good as about 400 m per pixel (Plate 7). The low latitudes are occupied by several distinct terrain types, which are overlain by a south polar cap that is thin enough near its edges to allow the morphology of the underlying terrain to show through.

The distribution of terrains is shown in Fig. 7.29. The three most widespread units are the high smooth plains (sh), hummocky plains (th), and the rather curious cantaloupe terrain (ct). It is fairly clear that the smooth plains unit overlies the other two, and in places the images show lobate scarps a few hundred metres to a couple of kilometres high, both at its margins and within it. This terrain has every appearance of having been emplaced as a series of large volume, highly viscous flows, such as the largely crystalline ammonia–water mixture suggested in one of the models for the flows on Ariel (section 6.4.3). With an abundance of methane, nitrogen, and probably some carbon monoxide available in

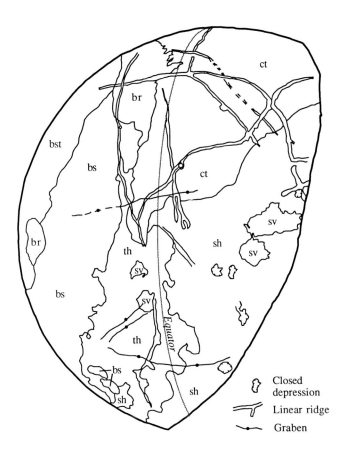

Fig. 7.29. Geological sketch map of terrains on Triton (same area as Plate 7). *sh* = high smooth plains; *sv* = smooth floor material; *th* = hummocky plains; *ct* = cantaloupe terrain; *bs*, *br* and *bst* are divisions of the polar cap (bright spotted, bright rugged and bright streaked). See text for explanation.

addition to ammonia to form intergranular fluids and a variety of partial melts, there is good scope for an especially wide range of volcanic phenomena during Triton's history. Several roughly circular depressions and chains of both rimmed and rimless pits are the probable sources of these outpourings.

A similar but low-lying unit is emplaced within both the high smooth plains and the hummocky plains, occurring in the floors of four flat-floored depressions up to about 200 km across. This has become known as smooth floor material (sv). The depressions are bounded by embayed escarpments, reminiscent of the escarpments in part of the plains terrain of Io (section 7.1.1), which in places are terraced where one segment of wall overlaps another (Fig. 7.30). It is not clear how much the present expression of the edges of these depressions is the result of constructional volcanic processes, how much was caused by caldera collapse, and how much is a product of the removal of overlying layers of some kind of erosional phenomenon, perhaps related to the loss of volatiles. In the centre of each depression is a rugged area near the limit of the image resolution, which appears to consist of a collection of pits and flows and to be the site of the most recent volcanic activity.

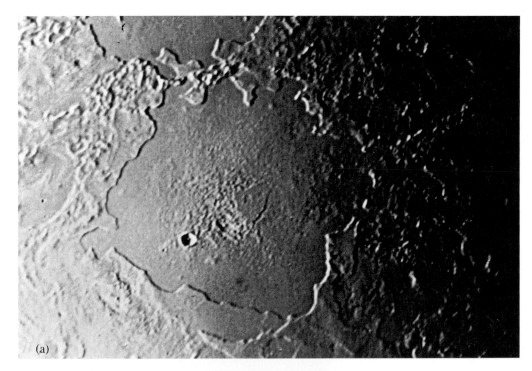

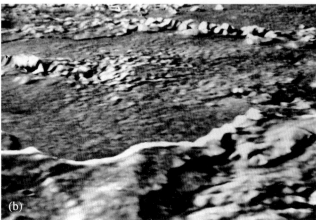

Fig. 7.30. (a) A smooth plains depression within Triton's high smooth plains unit. Note the overlapping terracing at the bottom of the image. The crater towards the lower left of the depression is 15 km across, and is probably an impact structure, whereas the finely textured area to its right may be evidence of volcanism. (b) A synthetic perspective view looking diagonally across the depression in (a) from the lower right.

The hummocky plains unit emerges from between the high smooth plains and the polar cap in Triton's leading hemisphere. It appears to be another flow-generated terrain containing a variety of domes, and one raised ridge, 700 km long, that looks like it was constructed by viscous flows at an eruptive fissure. A detailed view of hummocky terrain is shown in Fig. 7.31. The hummocky plains exhibit the greatest concentration of obvious impact craters on Triton, with a crater density comparable to that seen on the lunar maria.

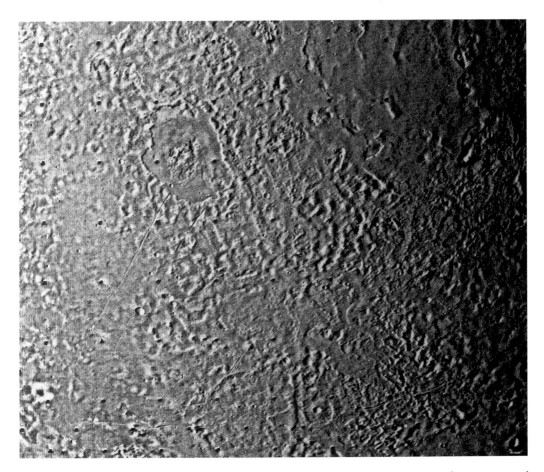

Fig. 7.31. Part of Triton's hummocky terrain, from the central lower parts of Fig. 7.29 and Plate 7. A ridge, apparently along a volcanic fissure, extends upwards from the lower edge of the view, skirting a smooth floored depression to the left of centre. Some nearly straight very narrow troughs (mapped as grabens in Fig. 7.29) can be seen in the lower part of the image. High smooth plains occupy the upper right portion of the image.

The crater density decreases away from the apex of orbital motion, which unfortunately tells us nothing of the origin of the impactors, because Triton's retrograde orbit makes its leading hemisphere take the full brunt of impacts both from debris orbiting Neptune and from material external to the Neptune system. It would be unwise to use comparisons of Triton's crater density with that of the lunar maria to imply an absolute age for the hummocky plains, but clearly this terrain is at least twice as old as the much more lightly cratered high smooth plains.

The other major terrain is of a type unlike any seen elsewhere in the solar system (Fig. 7.32). This consists of a dense patchwork of near-circular dimples, mostly either 25 or 5 km across, traversed by a network of ridges that appear to be viscous extrusions

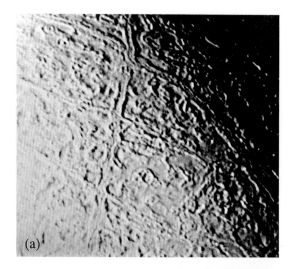

(a)

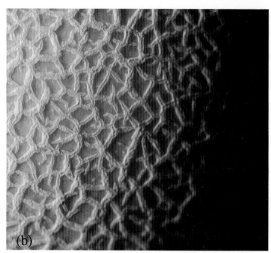

(b)

Fig. 7.32. (a) View of part of Triton's mysterious cantaloupe terrain, from the upper right of Plate 7. Several overlapping ridges may be made out, superimposed on a dense array of near-circular dimples. The image is about 450 km across. (b) Not part of Triton, but a close-up of the skin of a cantaloupe melon. The similarity in appearance is striking, although it casts little light on how Triton's cantaloupe terrain formed. Indeed, the comparison is doubly misleading, because the presence of methane and ammonia would almost certainly impart a greatly inferior flavour to Triton, compared to that of a real melon.

erupted along fault systems (most likely grabens). Because of the similarity in appearance to the skin of a well known variety of melon this has become known as the cantaloupe terrain. The origin of the dimples is a mystery. They may be a result of melting and collapse of the icy surface or they could be explosive craters like terrestrial maars. They may even be impact craters mantled or otherwise modified by later events, but they are all too similar in size to each other for this to seem likely. The evidence of superposition suggests that the cantaloupe terrain is the oldest unit on Triton and has been partly buried by both the smooth and the hummocky plains units, and even though it lies in the trailing hemisphere it is curious that no definite impact craters have been identified within it.

To judge from the pattern of linear ridges in the cantaloupe terrain and elsewhere, Triton's tectonics appear to be dominated by extension. These features seem to be either simple fissures or grabens that became flooded and indeed overfilled by viscous extrusions. In addition there are a few narrow troughs of the order of a kilometre across and up to 1000 km long that cut all units and are probably the most recent faults. These cause no offsets and are mapped as grabens on Fig. 7.29. Some workers have suggested the action of strike–slip tectonics near the boundary between the cantaloupe terrain and the smooth plains, on the basis of apparent offsets on the margins of one of the depressions, although no evidence of a surface fault-break is visible at the resolution of the available images.

In summary then, the terrain units on Triton are, in the plains areas at least, dominated by the extrusion of viscous, but often fairly extensive, flows. On the whole the surface appears to be relatively young, but despite the fascinating range of landforms there is no evidence of current activity. For this, we have to examine the polar cap.

7.4.3 *The polar cap*

Both Earth and Mars have polar caps, and the extent of these changes with the seasons. Frost migration on Callisto and Ganymede notwithstanding, the only known seasonal polar caps in the outer solar system appear to be on Triton, where the seasonal ice is probably mostly nitrogen, instead of water-ice (as on Earth) or water-ice and carbon dioxide (as on Mars). The *Voyager* images showed much of Triton's southern hemisphere to be occupied by a bright polar cap, with an albedo of around 0.88, whose ragged edge provides evidence that it may have been retreating at the time. To understand this it is necessary to appreciate that Neptune's (and Triton's) year is 165 Earth-years long and that a combination of Neptune's tilt and the inclination of Triton's orbit means that the sub-solar latitude on Triton varies from about 55° N to about 55° S during the year. Such a large range leads to the curious situation that Triton's tropics (55° N and S) lie further from the equator than its 'arctic' and 'antarctic' circles (35° N and S). At the time of the *Voyager-2* fly-by, Triton's southern hemisphere was about three-quarters of the way through its 40-year spring, and seasonal ice that had accumulated during the long darkness of the winter was being burned off by the Sun to be redeposited in the northern hemisphere.

This process offers a way to explain changes in Triton's spectrum. In 1979 this showed clear signs of methane, but during the following decade the methane absorption features became obscured and the spectrum began to show those traces of nitrogen that inspired the short-lived hopes of nitrogen oceans. In the light of the *Voyager* data, at least part of this spectral change can be attributed to the result of the migration of nitrogen from the southern polar cap and its redeposition in the northern hemisphere, which in 1979 still had regions of exposed methane deposits that had lost their nitrogen cover during the previous northern summer. Unfortunately, the change that this simple process could make in Triton's spectrum appears to be less than what was actually observed, and it is necessary to call for the deposition of fresh nitrogen frost over at least the fringes of the bright polar cap as well. This last point lends credence to models suggesting that the bright polar cap is actually rather stable in the long term, because it is formed of ices whose crystalline structure makes it highly reflective and therefore prevents the surface heating sufficiently for its full thickness to be completely lost by volatilization before the end of the summer. Seasonal changes, according to these models, are manifested mainly by the migration of a darker nitrogen frost that overlies the bright polar cap in winter and is volatilized and redeposited in the opposite hemisphere during the spring and summer.

Thus Triton's bright southern polar cap may not be the simple seasonal affair that it appears to be at first sight, in which case the seasonal changes are probably of considerable complexity. It is a pity that too little of the northern hemisphere was in sunlight at the time of the *Voyager* encounter to show whether or not there is a northern bright polar cap as well. In the event our observational evidence for polar processes on Triton comes from the southern hemisphere alone. On Fig. 7.29 three units are distinguished within the southern polar cap: bright rugged (br), bright spotted (bs) and bright streaked (bst). The bright rugged unit appears simply to be cantaloupe terrain with bright polar material resting in the hollows. If the bright polar cap shrinks as Triton's southern spring progresses into summer, it is likely that an increasing extent of cantaloupe terrain will be uncovered. Apparent evidence that the bright polar cap is in either long-term or seasonal retreat is provided by the bright spotted unit, where there is incomplete cover by polar ices. Here, roundish patches of the underlying surface show through, apparently on high ground, with thicker opaque ices in the low areas. This underlying unit is darker than most of what is exposed to the north, beyond the limits of the ice cap, and there is one curious area where an outlier of the bright spotted unit has been isolated by the retreat of the polar cap (Fig. 7.33).

The bright rugged and bright spotted units pass polewards into the bright streaked unit, which appears to be thick enough to obscure the underlying terrain completely. This unit, and parts of the spotted unit, is marked by streaks of dark material ranging from a few tens to about a hundred kilometres in length. Several of these can be seen on Plate 7, mostly oriented towards the north-east, except within 10° of the pole, where they point in all directions and may cross one another. Typically, a streak fans out slightly down-length, and usually has a dark patch a few kilometres across at its head. The dark colour of the streaks and dark spots may be due to complex organic polymers produced from methane by the action of charged particles trapped in Neptune's magnetic field, or by cosmic rays and ultraviolet photons.

Fig. 7.33. Isolated areas of Triton's bright spotted unit (from the lower part of Plate 7), apparently left behind by the retreating polar cap. The mushroom-shaped dark patch in the foreground is about 100 km across.

If they lie on top of the seasonal ice these streaks must be ephemeral features, but if the bright polar cap is stable then many may date from previous years. They look similar to streaks formed by deposits of windblown dust on Mars, and it has been calculated that on Triton grains 5 μm in diameter or less could be carried in suspension by winds of about 10 m s^{-1}, the thinness of the atmosphere being compensated for by Triton's low surface gravity (0.78 m s^{-2}, less than a tenth of Earth's gravity). Such a speed is force 5 (a fresh breeze) on the terrestrial Beaufort scale, and is well within the bounds of possibility on Triton.

A prolonged discussion of Triton's meteorology would be both premature and beyond the scope of this book, but it is instructive to make a few comparisons with other planets. On Mars the atmosphere contains about three to five times as much mass as the seasonal polar caps, but on Triton the atmosphere has only about a hundredth of the mass of its observed seasonal cap (depending on how much of the cap is actually seasonal). This means that Triton's atmosphere can be regarded merely as the polar material in transit; whatever is sublimed from the south polar cap as it shrinks must be transported northwards to be deposited in the growing north polar cap at a roughly equivalent rate, and if virtually the whole south polar cap is to be removed before the end of the summer then the time each nitrogen molecule spends in transit can be no more than a small fraction of a Triton year. Consequently, the atmospheric pressure is likely to be very unstable, and it would be feasible for it to increase or decrease by a factor of ten or more from the 16 μbar value measured at the time of the *Voyager* encounter.

As the south polar cap sublimes the local atmospheric pressure must increase, and winds will develop that cause a net transport of vapour away from the region. However, on a rotating body such as Triton such winds cannot simply blow directly from south to north. Instead, there is likely to be (at least near ground level) a polar anticyclone with winds blowing north-eastwards at about 10 m s^{-1} in a spiral pattern round the pole. This deflection of the winds to the right in the southern hemisphere is the opposite of what would be found on any other body in the solar system, because, as a result of its retrograde orbit and captured rotation, Triton rotates from east to west, instead of from west to east.

Thus the generally north-east orientation of the dark streaks on the south polar cap is compatible with anticipated wind directions, and the streaks themselves can be explained if the black patches usually found near their heads are sites where suitable particles can be picked up. Although the origin and composition of this material is uncertain, it is clear that if these particles are picked up from the ground by the wind then they must display a remarkably low degree of cohesiveness. On Earth fine-grained particles tend to stick together because of a thin film of water on their surface and other molecular forces, but if such cohesion were to occur between particles on Triton it would be impossible for the anticipated 10 m s^{-1} winds to pick them up.

7.4.4 Plumes

One way to side-step this problem is to have the particles thrown up into the air by some kind of eruptive process, because this would probably separate even cohesive grains, which, once airborne, would be at the mercy of the winds. At least two such eruptions on the polar cap were imaged during the *Voyager-2* fly-by, one of which is shown in Fig. 7.34. Each of these was powerful enough to send a jet of dark material vertically upwards to an altitude of 8 km, at which height it bent sharply over to become a dark horizontal cloud extending westwards for over 100 km. Like certain major volcanic eruption plumes on Earth, it appears that Triton's plumes cease to rise when they reach an altitude at which they are no longer buoyant; probably in this case the 8 km limit marks the tropopause, above which the temperature of the atmosphere starts to increase.

The fast westward-blowing winds at high altitudes required to account for the orientation and length of the horizontal part of the plumes can actually be made to tally with models calling for slower winds blowing north-eastwards at ground level. However, particles falling-out from such a plume would produce an elongated dark deposit over the surface ice with an orientation quite different from that commonly shown by the dark streaks. Possible explanations are that most of the dark streaks come from dust picked up directly from the surface by the wind and that only a few are due to plumes of the type seen on the images, or that there were changes in Triton's weather pattern earlier in the spring before *Voyager-2* arrived.

As to the cause of the plumes, there are probably as many explanations as there are scientists who have studied them. Their source at the polar cap argues strongly that nitrogen is the driving gas, although methane remains a possibility. The energy source could be from within Triton, either in the form of the background heat-flow or from the intrusion of comparatively warm icy lavas below the polar ice, either of which could vaporize this ice to liberate nitrogen gas. Such mechanisms would be closely analogous to

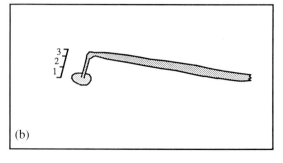

Fig. 7.34. (a) *Voyager-2* image of a dark plume erupting over Triton's south polar cap. Its source is a dark spot on the surface, and it rises to 8 km before dispersing downwind at a constant altitude. Triton is the only body in the solar system, other than Io and the Earth, where eruption processes have definitely been seen in action. In this case, the activity may be more geyser-like than volcanic. (b) Sketch interpretation of the plume shown in (a). In the height interval marked 1, the plume is rising through a combination of its erupted momentum and buoyancy forces; in the height interval marked 2 it is rising mainly through buoyancy, because it is warmer than the surrounding atmosphere. At the tropopause (height 3) the atmospheric temperature gradient reverses and the plume becomes neutrally buoyant, so it ceases to rise. The wind at this altitude then streaks the plume out over a length of some 100 km.

geysers on the Earth, where volcanically enhanced heat-flow causes the flash vaporization of water to steam. An alternative, suggested by the position of the two observed plumes close to the latitude where the Sun was overhead at midday, is that solar energy is responsible. This model is illustrated in Fig. 7.35. It shows a layer of nitrogen-ice about 2 m thick overlying a dark substrate. The ice is transparent and allows most of the sunlight to penetrate to the dark layer where it is absorbed and converted to heat. A rise of only 10 K above the surface temperature of 37 K would cause the vapour pressure of nitrogen to rise about a hundred-fold, so the conduction of heat into the ice from the solar-heated dark substrate could allow a large volume of pressurized nitrogen vapour to build up, probably in interconnecting fissures rather than in a vast blister as implied in the simple sketch in the figure. Eventually the solid nitrogen cap would rupture, allowing gas to escape rapidly, accelerating as it decompressed and picking up dark particles and other material from below and within the exit nozzle.

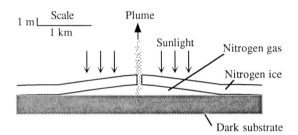

Fig. 7.35. A possible eruption mechanism for Triton's plumes, here interpreted as a geyser-like eruption driven by the pressure of nitrogen gas escaping from beneath a 'greenhouse' layer of nitrogen-ice. Note the vertical exaggeration.

It would certainly be fascinating to return to Triton in the southern hemisphere summer, to see what remains of the southern polar cap, whether any of the dark streaks have survived the removal of the ice, and whether any plumes are active. For the foreseeable future, however, we shall have to make do with what *Voyager-2* showed us. Twenty years ago most of us would probably have settled for one look, but now it is clear that Triton is one of the most fascinating worlds in the entire solar system, which many would now place high on their repeat wish-list.

8

Unseen worlds

What is this world? what asketh men to have?
Chaucer, *The Knightes Tale*

We have now discussed all the major satellites in the outer solar system of which the existing images permit us make direct geological inferences. There remain just three more bodies that we need to consider to make our survey of this geology of the outer solar system complete. These are Titan, the largest satellite of Saturn, and the remarkable double system of Pluto and Charon. The former is wreathed in an opaque atmosphere, and the latter pair has not yet been visited by a space probe.

8.1 Titan

Titan was discovered in 1655 by the Dutch scientist Christiaan Huygens. Although it was the next satellite to be found after the galilean moons of Jupiter, it has kept many of its secrets up to the present day. Titan's methane-rich atmosphere, which hides its surface, was discovered in 1944 by Gerard Kuiper (another Dutchman, although working in America) by means of spectroscopic techniques, and several other constituents were identified prior to the *Voyager* encounters.

Titan is a large world, in the same class as Ganymede and Callisto in terms of both size and density (Table 1.1). The radius of its solid surface is about 2575 km, only 60 km less than that of Ganymede. As its atmosphere is opaque to an altitude of some 200 km, this thickness is sometimes (rather unreasonably) added to the solid radius to give Titan the honour of being the largest satellite in the solar system.

As the only solid body beyond Mars with a substantial atmosphere, and an organic-rich one at that, Titan was a target of prime importance for *Voyager* investigation. *Voyager-1* was sent on a trajectory that took it through Titan's Earth and Sun occultation zones; the former so that radio signals from the probe could pass through Titan's atmosphere *en route* to Earth, and the latter so that the atmosphere could be studied by looking through it towards the Sun. It was these constraints that made it necessary for the probe to continue over Saturn's south polar region, whence the gravitational sling-shot effect flung it northwards out of the plane of the ecliptic, thus precluding any further planetary encounters.

During the hours before closest approach, *Voyager-1* imaged most of Titan's illuminated disc with a resolution better than 2 km per pixel, and half of it better than 650 m per pixel, but no breaks in the cloud were to be seen. Instead, Titan was shown to exhibit a completely opaque red aerosol-rich layer 200 km above the ground surface, consisting of microscopic droplets of hydrocarbons, with several haze layers above (Plate 8). The *Voyager* images show the aerosol layer to be darkest in a hood around the north pole and somewhat brighter in the southern hemisphere than the northern (probably a seasonal effect); but on the whole it is bland in the extreme, with just a few east–west markings that show up under extreme contrast enhancement.

Voyager-1's gravitational interaction with Titan gave the first accurate determination of its mass, and the radio occultation data measured its radius to within an error of about 2 km. These data suggest a rock to ice ratio of about 52:48. However, because we lack the crucial evidence of Titan's surface morphology that would have been provided by images of its surface, we are unable to say very much about its geological history. Extreme models for its present structure are shown in Fig. 8.1; accretional and radiogenic heat were probably sufficient to allow Titan to differentiate to some extent at least, especially if the ice was originally volatile-rich as implied by the thickness and composition of the atmosphere.

Both *Voyagers* probed the nature of Titan's atmosphere with a variety of instruments that spanned the spectrum from the ultraviolet to radio wavelengths. The attenuation of the radio signal from *Voyager-1* as it passed behind Titan suggests that the surface temperature and atmospheric pressure are 94 K and 1.5 bar respectively. The latter value means that Titan's ground-level atmospheric pressure is 50 per cent greater than at sea-level on Earth. Ultraviolet and infra-red spectrometer data indicate that Titan's atmosphere consists mostly of nitrogen with between 2 and 10 per cent methane. Other species that have been detected are hydrogen (about 0.2 per cent) and smaller traces of other, mostly organic, gases including ethane (C_2H_6), propane (C_3H_8), acetylene (C_2H_2), hydrogen cyanide (HCN), and carbon monoxide (CO).

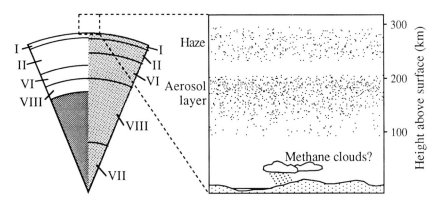

Fig. 8.1. Extreme alternative models for the internal structure of Titan, assuming a 52:48 (by mass) rock to ice ratio, showing the phases of water-ice that would be stable. On the left, a fully-differentiated model, with all the silicates segregated into the core, on the right an undifferentiated model, which is less likely. The box shows the structure of the atmosphere.

Argon has not been detected, but some workers have suggested that several per cent of argon is required in the atmosphere to increase the average molecular weight of the mixture to the value implied by the radio occultation data. The matter of the argon is unresolved, but it seems clear that another of the inert gases, neon, cannot be present in any amount, because this would decrease the average molecular weight below what is compatible with the present data. This has important implications for the origin of Titan's atmosphere, which ought to have roughly equal abundances of nitrogen and neon had it been captured directly from the solar nebula or the proto-Saturn nebula. The lack of any large fraction of neon shows that the atmosphere almost certainly formed by degassing out of the ice and rock of which Titan is composed.

The original ice would have been a clathrate, with methane, nitrogen (perhaps in the form of ammonia), and argon (if present) held in voids within the water-ice lattice. After degassing, which was probably associated with internal differentiation, solar ultraviolet photons would have begun to break down any ammonia to nitrogen (N_2), which is heavy and has remained in the atmosphere today, and hydrogen (H_2), which is light and has almost all escaped. In the meantime, if ammonia ever had the chance to build up to a significant concentration in the atmosphere, it would probably have acted as a 'greenhouse' gas, trapping solar heat and raising the surface temperature by tens or even hundreds of degrees. Apart from breaking down the ammonia, ultraviolet light has probably always played an important role in modifying Titan's atmosphere, by removing hydrogen atoms from methane and thus initiating the assembly of the larger organic molecules and cyanide compounds found today.

Although we have no direct information on what goes on below the aerosol layer, the weird conditions of Titan's atmosphere ought to allow both methane and ethane to condense. In particular, it has been calculated that, over time, enough ethane has probably been made in the upper atmosphere to have drizzled downwards and formed a liquid layer around 1 km deep. Just as for Triton, hopes have been raised of finding another world with oceans. These ambitions appear to have been dashed, temporarily at least, by an Earth-based experiment. In June 1989, using antennae that were needed immediately afterwards for communications with *Voyager-2* during its approach to Neptune, a radar beam was bounced off Titan's surface. The strength of the returned signal was ten times that expected from a global ocean even as shallow as 200 m, and was more like an echo from a dry, rough surface. Thus at present the global ocean appears to be ruled out, although smaller seas and lakes could still be there in abundance.

Apart from the background drizzle of ethane originating at high levels, there may also be clouds of methane condensing lower down, giving rise to either methane snow or methane rain. So, if we could see Titan's surface, we might well find it to be sculpted by the action of winds, flowing liquids, and wave action, resulting in an enormous range of erosional and depositional processes, similar to those with which we are familiar on Earth, and yet probably different in a surprising number of ways. As described in the final chapter, radar images from the planned *Cassini* space probe may give us our first good view of all this early in the next century.

Titan's atmosphere is strikingly unlike that of the Earth, but it is the most similar atmosphere known to that of the Earth about 4 billion years ago when (in the absence of free oxygen) carbon monoxide, carbon dioxide, and nitrogen combined to build the first nucleic acid molecules that led to the development of life. For this reason, and because we have yet to learn anything about the tectonic, volcanic, and sedimentary evolution of its surface, Titan is the solar system's most important and intriguing unmapped target for future exploration.

8.2 Pluto and Charon

For the first forty years or so after its discovery in 1930 Pluto, the ninth planet, was regarded as a solar system misfit: too small to be a gas giant and too far from the Sun to be a terrestrial planet—fascinating, no doubt, but no place for a geologist. Pluto appears so small in an Earth-based telescope that the diameter of its disc cannot be measured with any accuracy, and until the mid-1950s the best estimates were that it was about half the size of the Earth, or perhaps slightly bigger. This could hardly be the ten-Earth masses body required to cause the perturbations in Neptune's orbit that led to the tracking down of Pluto in the first place, which now appears to have been a piece of good luck as much as anything.

The accepted size of Pluto began a further dive when, in 1976, solid methane was detected in Pluto's spectrum. It became clear that the surface is icy and comparatively highly reflective, calling for a small high-albedo body rather than a larger low-albedo body. Two years later, high-resolution telescopic photographs revealed the presence of a satellite in orbit around Pluto. Named Charon, after the ferryman who in mythology conveyed the dead into the underworld (the domain of the god Pluto), this satellite soon began to provide important information. The dimensions and period of its orbit enabled the combined mass of the Pluto–Charon system to be calculated showing that Pluto must be far less massive than the Earth, and indeed is outdone even by Triton. Some of the best images we have of Pluto and Charon are shown in Fig. 8.2.

Charon's orbit about Pluto is inclined at 94° to Pluto's orbit about the Sun, so technically it is retrograde. Tidal interactions make it virtually certain that the plane of Charon's orbit coincides with the plane of Pluto's equator, so the 94° inclination implies that Pluto itself is tipped on its side. Furthermore Charon's orbital period of 6.4 days is the same as the rotation period of Pluto that can be deduced from periodic variations in its brightness. Thus Pluto always keeps the same face turned towards Charon, and presumably Charon's rotation is tidally locked to Pluto in a mutual fashion.

The (relative) flood of information about Pluto and Charon was boosted in the years 1985–1990 when there were many occasions when the line of sight from Earth passed through the plane of Charon's orbit about Pluto, allowing Charon to pass directly behind and in front of Pluto. By timing and measuring the changes in combined brightness during these events the relative sizes of Pluto and Charon could be calculated, and inferences could even be made about the general distribution of surface albedo patterns. This series of occultations was a definite stroke of good luck, because such a spell of favourable

Fig. 8.2. Pluto and Charon imaged from Earth in 1990. Top left, the view from a ground based telescope in Hawaii, showing blurring of the image due to the Earth's atmosphere. Top right, the view from the Hubble Space Telescope in orbit; the faint rings are artefacts of the enhancement process. The positions and relative sizes of Pluto and Charon are shown to scale in the sketch at the bottom.

alignments occurs only twice in Pluto's 248-year orbit about the Sun. Had Charon been discovered just twelve years later we would have had to wait a very long time before we were once again in a position to gather such precise information.

The modern estimates for the sizes, masses, and density of Pluto and Charon are given in Table 1.1. Their relative masses probably differ by a factor of only about ten, making Charon much the most massive satellite with respect to its planet. Previously, this honour went to our own Moon. Charon orbits closer to its planet than any other satellite in the solar system (large or small), and seen from the surface of Pluto it would have an angular diameter of almost four degrees, eight times the apparent diameter of the Moon in the Earth's sky. The development of such a close relationship is likely to have been associated with considerable tidal heating, and it is probable that both bodies have had the opportunity to become fully differentiated. Charon appears to have a lower albedo than Pluto, and, so far as can be told, its spectrum is lacking in methane features and appears to show water-ice only. If they are differentiated, Pluto and Charon should each have a sizeable rocky core overlain by a mantle of water-ice and topped, probably in the case of Pluto only, by methane (Fig. 8.3).

The similarities in size, density, and composition between Pluto and Triton now seem rather striking. Further similarities emerge when we consider Pluto's atmosphere. After the discovery of methane in Pluto's spectrum it was suggested that some of the absorption features were due to methane in the gaseous state rather than in surface ice. The presence of an atmosphere around Pluto was proved when, in June 1988, the passage of Pluto in front of a star was observed for the first time. The light from the star began to fade

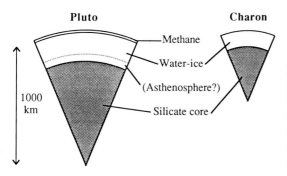

Fig. 8.3. Possible internal structures of Pluto and Charon. The best estimate for Pluto's bulk composition is that it has 68–80 per cent silicates, with the balance made up of ices. Where water-ice is shown, this could include several per cent of nitrogen and carbon monoxide.

gradually as it was attenuated by Pluto's atmosphere, the results suggesting an atmospheric pressure less than a hundred thousandth of the Earth's, but similar to Triton's. Apart from methane, heavier gases such as nitrogen and carbon monoxide may well be present.

It is particularly tempting to suggest nitrogen, as this is the major component of the atmosphere of Triton. Like Triton, Pluto experiences severe seasonal changes. Apart from Pluto's 90° axial tilt, its orbit is highly eccentric, so that during its year its distance from the Sun varies from 7.4 billion kilometres (at aphelion) to 4.5 billion kilometres (at perihelion), producing extreme variations in the global mean temperature. Pluto reached perihelion in 1989 (being closer to the Sun than Neptune during the years 1979–1999) and it is probable that its atmosphere is a temporary phenomenon, being volatilized from the surface only during the sixty or so years closest to perihelion. This could explain Pluto's high albedo, because, whereas methane ices elsewhere (such as the water–methane clathrate suggested for the satellites of Uranus) appear to darken with age, the methane ice on Pluto is reformed each Pluto-year.

On the other hand, it has been argued that during the half of its year when Pluto is closest to the Sun, methane will escape from its atmosphere to space at a significant rate. To maintain a long-term steady state, the atmospheric methane would need to be replenished by a net loss of about a millimetre of methane-ice from the surface per orbit, implying a loss of approaching 10 km of surface methane over the lifetime of the solar system. This has been used as a powerful argument in favour of Pluto being fully differentiated, because if there were even a tiny fraction of particles of rock or water-ice mixed with the crustal methane layer these should, over only a few Pluto-years, form a surface lag deposit thick enough to insulate the underlying methane from solar heating. Thus, according to this logic, Pluto appears to have an outer layer of rather pure methane that, if it was formed early in the lifetime of the solar system, must have originally exceeded about 10 km in thickness. This is not far short of the total amount of methane expected in a body of Pluto's mass on cosmochemical grounds. Charon, with its lower gravity, has evidently lost all the methane it ever had available near the surface.

If the rate of loss of methane from the surface of Pluto has just been correctly described, then Pluto may be a rather featureless world. Primordial craters less than about 5–10 km deep will have been erased, together with any ancient topographic features resulting from tectonic and other processes. In contrast, Charon, if it finished losing its methane early on, will have a much fuller cratering record preserved on its surface.

Like Triton, Pluto's relatively low abundance of water compared to rock argues for accretion in the cold outer region of the solar nebula. However, whereas Triton was subsequently captured by a major planet, Pluto avoided this fate. Instead Pluto has wandered until its orbital period has become locked in a stable 3:2 orbital resonance with Neptune's orbit about the Sun. Whether Pluto, and indeed Charon, experienced a heating event comparable to that responsible for the variety of relatively young terrain types on Triton remains unknown. Tidal heating due to interaction with Neptune was probably negligible, but tidal interactions between Pluto and Charon could have been a cause of considerable heating. Differentiation within Pluto, if not Charon, may have been aided by a significant amount of radiogenic heat.

How Pluto and Charon came to be together as a pair is mysterious. Separate origins followed by gravitational capture would require a seemingly too improbable sequence of events. It is possible that they could have accreted as a double planet, as in a currently out-of-favour model for the Earth–Moon system. Alternatively, Charon could have accreted from the debris resulting from a collision between Pluto and a major planetesimal. This debris would consist of the remains of the planetesimal and parts of Pluto's outer layers, along the lines of the currently favoured model for the Moon's origin. The presently modelled densities of Pluto and Charon are so close that the double planet model is the more compelling, but the data are poorly constrained and the matter is unlikely to move much closer to solution until we have much better surface spectra and detailed images.

We shall have to wait for a while for the latter, because it would take about 14 years for a probe to reach Pluto, even if a favourable opportunity at the turn of the century is seized upon. Unfortunately, at the time of writing, no space agency has been able to commit the (relatively modest) sum of $200 million that the launch of such a mission would require. We shall turn in the final chapter to a summary of the most important things we need to find out to understand the geology of our companion worlds in the outer solar system, and the missions that are planned that should go at least some way towards achieving this.

9

What next?

Come my friends,
'Tis not too late to seek a newer world
Tennyson, Ulysses

It is in the nature of science always to want to know more. This trait is certainly no less marked in the planetary sciences than in other disciplines. Ten years after the first Moon landing (in 1969), we still had very little information on the moons of the outer planets. Today we have maps of large parts of most of them and can speculate about their composition and evolution using the landforms and spectra of their surfaces, supported by cosmochemical models. However, there is still a great deal that we need to find out. Many of the present interpretations are probably flawed in detail, and some are certain to be downright wrong.

9.1 The missing data

Clearly, the satellites of the outer planets have a lot to tell about the variety of styles in which planetary bodies with different thicknesses of lithosphere and different histories of internal heating can evolve. What then do we most need to determine if we are to understand the origin and evolution of the outer planet satellites, and to take advantage of the light they can cast on planetary bodies in general? As is apparent in the preceding chapters, each one of these worlds has its own characteristics, so we are faced with trying to understand the histories of at least nineteen individual bodies. Generalizations are unwise, and it is dangerous to extrapolate from one body to another.

One thing that would help enormously would be if we could analyse samples collected from the surfaces of some of the outer planet satellites. In particular, the evidence for methane and ammonia in many of the satellites of Saturn and Uranus is largely circumstantial, and furthermore we have only a poor concept of the relative importances of silicates and carbonaceous material in the dense fractions of these bodies. Sampling of surface material is probably several decades away (with the possible exception of Titan), and the most we can hope for is better remote analysis using improved surface spectra obtained by fly-by probes and Earth-based observations. One example of a major outstanding mystery is that of the composition of the dark material on Iapetus, and whether its dark-floored craters contain the same material as that blanketing the dark hemisphere. If this could be solved it would tell us a considerable amount about both

Iapetus's surface and its interior, and the possibility of a volcanic origin for the source of the dark material.

To get a good idea of the internal structure of a planetary body it is necessary to deploy seismometers on its surface. By examining the nature and relative timing of vibrations that have propagated from the sources of seismic disturbance, such as 'earthquakes' and meteorite impacts, internal phase changes and other interfaces between layers of different density can be located, densities can be inferred, and the presence of liquid or partially molten zones can be deduced from the attenuation or non-transmission of shearing waves (as opposed to pressure waves). The possibility of these sorts of experiments in the outer solar system seems a long way off at present (so far, apart from on Earth, they have been achieved only on the Moon and, very crudely, on Mars). A less ambitious option would be to use detailed measurements of a satellite's gravity field, derived from perturbations of the trajectory of a probe in a close fly-by, to infer internal density variations.

As for the imaging of satellite surfaces, we need fuller coverage at the present best resolutions to prevent us making inappropriate generalizations about a whole body based on good pictures of only a fraction of its surface, and also some images with finer resolution to enable smaller details to be picked out. The way our perception of Mars as a Moon-like cratered planet, based on the low resolution and patchy coverage obtained by *Mariner-4* in 1965, was changed by the better and more extensive mapping by *Mariner-9* (1971) and the two *Viking* missions (1976) to that of a varied world with large volcanoes, canyons, and tracts of sand dunes is an object lesson in this respect. In addition to obtaining fuller coverage at higher resolutions, it would be preferable to obtain pictures of coincident areas seen from different directions. This would provide stereoscopic images that would enable the surface topography to be mapped, overcoming many of the limitations of shadow-length measurements and shape-from-shading techniques, which are all we have available at present.

By way of example, some of the issues that could be settled by more extensive and improved imaging are as follows. We do not know whether the asymmetric distribution of mountainous terrain on Io (and the crustal asymmetry that this implies) is real, or whether it is just an artefact of the restricted resolution coverage provided by the *Voyager* probes. Furthermore, re-imaging of Io would tell us about the extent of volcanic activity in the interim, by revealing any new plume deposits, new lava flows, and new hot-spots, as well as telling us about the persistence or otherwise of activity at the volcanic centres that are already known, and helping to settle the controversy over sulphur versus silicate volcanism. In the case of Europa, more extensive and higher resolution imaging is required to reveal the presence of zones of lithospheric convergence to balance the extension implied by many of the dark bands, to help us date the surface terrains by means of the density of craters less than 1 km in diameter, and possibly to detect active plumes arising from explosive water and ice eruptions. Similarly improved coverage would enable us to complete the global tectonic map of Enceladus and to assess the extent, if any, of present-day activity. Rhea is another world where the *Voyager* coverage is frustrating. In this case the anti-Saturn hemisphere was poorly imaged (50 km resolution), just good enough to show pale wispy markings that might be the former sites of major tectonic activity, on a world where models of its thermal evolution suggest that we should expect some signs of geological events to have survived. Titan is in a class of its own: only radar images would be capable of revealing details of its tectonic and volcanic history, and its surface morphology may contain evidence of changes in its climate linked to the 'greenhouse' effect and other factors.

We are also lacking complete global views of the satellites of Uranus and Neptune. Among these Titania's evident tectonic history makes the low resolution of the images of it particularly frustrating. There is no immediate prospect of remedying this situation, but missions are *en route* or planned to both Jupiter and Saturn that should greatly improve our understanding of these planets and their attendant satellites. These are now briefly described.

9.2 *Galileo*

Galileo, the most ambitious mission to the outer solar system yet, is due to arrive at Jupiter late in 1995, and will be the first spacecraft to go into orbit round a gas giant planet. Named in honour of the discoverer of Jupiter's main satellites, *Galileo* was much delayed by development problems and then the suspension of Space Shuttle operations following the accident that destroyed the Space Shuttle *Challenger* in 1986. *Galileo* was eventually sent on its way from the cargo bay of a Space Shuttle in October 1989. It was boosted out of Earth orbit by a low-thrust upper-stage rocket in place of the more powerful type of rocket that would have been used before the *Challenger* disaster, but which was then ruled to be too hazardous a cargo. As a result, the only way of giving *Galileo* sufficient velocity to reach Jupiter was to use gravitational assists, analogous to the gravitational sling-shot effect used to accelerate the *Voyager* probes past the gas giants. As shown in Fig. 9.1, *Galileo*'s trajectory took it first past Venus (at an altitude of 19 000 km) and then on to even lower altitude swings past the Earth in December 1990 (960 km) and December 1992 (300 km), with an opportunity for a fly-by of an asteroid in between. On arrival in the Jupiter system, *Galileo* should make a close pass of Io, about 1000 km above its surface (twenty times closer than *Voyager-1* reached), and use a kind of reverse gravitational sling-shot effect, assisted by rocket thrusts, to slow it down enough to be captured into orbit around Jupiter.

Galileo's instrument package is more sophisticated than *Voyager*'s. The imaging system is more sensitive, and there is a near-infra-red mapping spectrometer that will be of considerable help in determining the compositions of satellite surfaces, and of Jupiter's atmosphere. There are several instruments to probe the magnetic environment and to characterize the charged particles better, including those arising from volcanic activity on Io.

Galileo will stay outside the dangerous radiation belts around Jupiter, except for the initial close approach taking it within the orbit of Io that is necessary for capture. As a result the inbound close pass of Io will be the only opportunity for detailed imaging of it, down to 20 m resolution (about as good as that achieved by publicly available satellite

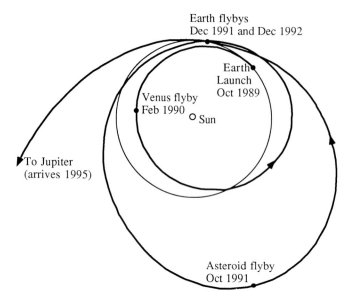

Fig. 9.1. The Venus–Earth–Earth gravity-assist trajectory used to speed Galileo outwards for its 1995 arrival in the Jupiter system. It recorded the first ever close-up images of an asteroid, 951 Gaspra, in October 1991.

images of the Earth), but covering only a limited area. However, during the remaining 22 months of the projected mission, *Galileo* will be able to monitor Io with a resolution of 10–20 km, which should be adequate to keep track of changes in volcanic activity, especially plumes and hot-spots.

Once in orbit around Jupiter, *Galileo* will follow a complex trajectory involving a series of encounters with Europa, Ganymede, and Callisto, often at less than 1000 km, that should enable parts of their surfaces to be imaged with a resolution as small as 20 m, and offering near-global coverage at about 1 km resolution. This ought to show fascinating detail of the morphology of such features as Ganymede's grooves and Europa's cycloid ridges. The repeated close passes of the three outer galilean satellites should offer considerable scope for determining their internal mass distributions, and there will be scope for a more limited study of this kind during the single Io encounter.

Jupiter's atmospheric circulation will be studied, via imaging and other experiments, and its composition will be determined both by remote sensing experiments and by means of a 118 kg probe that will descend through Jupiter's clouds by parachute. These analyses may cast new light on the chemistry of the solar nebula in the Jupiter region, upon which models for the composition of Jupiter's satellites depend.

9.3 *Cassini*

If all goes well, *Galileo* will be followed by an equally exciting mission to the Saturn system, as a joint enterprise between NASA and the European Space Agency. Named after the discoverer of the widest gap in Saturn's rings, *Cassini* is scheduled for launch in

April 1996 and for arrival at Saturn in December 2002. It should make gravity-assist fly-bys of the Earth in 1998 and of Jupiter in 2000, and will probably fly within 4000 km of at least one asteroid *en route*.

Once at Saturn, *Cassini* will follow a complex orbital pattern (Fig. 9.2), similar to that of *Galileo* at Jupiter. The prime target of the mission is Titan, which will be investigated by means of a probe that will perform atmospheric analyses during a parachute descent, and may survive long enough to analyse the surface material. The orbiter will carry an imaging radar to map surface features on Titan, and other microwave instruments to study the physical properties of its surface material. The potential fruits of such an instrument package were demonstrated by the *Magellan* probe, which produced spectacular and detailed radar images of the surface of Venus in 1990 and 1991.

Another, more conventional, imaging system carried by *Cassini* will cover Saturn itself and its other satellites. Because of limited fuel for manoeuvring and because there is less opportunity to play gravitational ping-pong amongst Saturn's low-mass satellites than in the Jupiter system, *Cassini* will probably make only four satellite passes at less than 1000 km during its nominal four-year orbital mission—twice past Iapetus and once each past Enceladus and Dione—but there will be several more distant encounters with these and other satellites that should, in total, offer a considerable advance on what we know from the *Voyager* data.

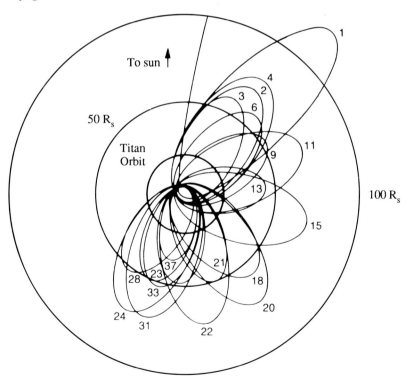

Fig. 9.2. The possible orbital pattern of *Cassini* upon arrival in the Saturn system. The numbers refer to successive orbits. R_s is the radius of Saturn (60 000 km).

9.4 Boldly going . . .

Thus, despite the perennial cut-backs, we are likely to continue to amass a considerable volume of new data on several of the outer planet satellites over the coming ten years or so. However, there seems to be little prospect of a prompt return to Uranus and beyond, so the mysteries of Triton's climatic changes and the volcanism and tectonism on Miranda, Ariel, and Titania will remain unresolved for now. Costs, potential scientific returns, and the increased risk of malfunction during a mission of long duration must all be weighed up when assigning priorities for exploration. This means that the delights of the outer solar system have to be balanced against what could be achieved by the more detailed exploration of objects nearer to home, such as the Moon, Mars, and the asteroids, where there may even be economic pay-backs in the medium-to-long term.

As suggested at the end of Chapter 8, if launched in the year 2001 a low cost Pluto–Charon mission could take advantage of Jupiter's favourable position for a gravitational sling-shot manoeuvre, but this may well remain just a pipe-dream. If so, we have passed the end of what others before me have called 'the golden age of solar system exploration', during which we gained our first views of so many worlds never before seen in any detail. History may well count the dozen years between *Voyager-1*'s Jupiter encounter in 1979 and *Voyager-2*'s Neptune–Triton fly-by in 1989, when the surfaces of no fewer than sixteen satellites greater than about 200 km in radius were charted for the first time, as the crowning achievement of this era. For planetary geologists, there is unlikely to be such an eye-opening episode again until such time as we see our first detailed images of the planets of another star. But that, so to speak, will be another story.

Appendix

Names of surface features

The names of features on the outer planet satellites have been agreed by the International Astronomical Union according to the following conventions:

Satellites of Jupiter

Io	Gods and heroes associated with fire, sun, thunder, or volcanoes; also people and places associated with the Io myth.
Europa	European (Celtic) gods and heroes, places in ancient Egypt, people and places associated with the Europa myth.
Ganymede	Gods, heroes, and places from ancient Egypt and the fertile crescent.
Callisto	People and places from northern myths.

Satellites of Saturn

Mimas	People and places from Sir Thomas Malory's *Le Morte d'Arthur*.
Enceladus	People and places from Sir Richard Burton's version of the *Arabian Nights*.
Tethys	People and places from Homer's *Odyssey*.
Dione	People and places from Virgil's *Aeneid*.
Rhea	People and places from African, Asiatic, and South American creation myths.
Titan	Not yet mapped.
Iapetus	People and places from the medieval French epic *Chanson de Roland* (Song of Roland).

Satellites of Uranus

Miranda	Humans from Shakespeare's *The Tempest*; Shakespearean place names.
Ariel	Bright (good) spirits from worldwide mythologies.
Umbriel	Dark (evil) spirits from worldwide mythologies.
Titania	Minor Shakespearean female characters.
Oberon	Shakespearean heroes.

Satellites of Neptune

Triton	Aquatic gods and places.
Pluto and Charon	Not yet mapped.

Glossary

accretion The formation and growth of primitive planetary bodies by collision with other bodies and gravitational sweeping up of gas and dust.

accretional heating Heating of a planetary body as it grows by the kinetic energy liberated when smaller objects collide with it. Most of this heat is generated near the temporary surface, but can be carried inwards by conduction and **convection**.

aerosol Liquid droplets small enough to be supported in an atmosphere. Titan's atmosphere is opaque because of aerosols of ethane and other, more complex, organic molecules.

albedo A measure of the brightness or reflectivity of a surface. An albedo of 1 means that all the incident light is reflected by a surface, so that it appears white. (In this book, the values of albedo quoted are of 'normal' or 'geometric' albedo calculated for light falling perpendicularly on the surface, for small areas, or along the line of sight, for spheres). For an icy moon, an albedo of nearly 1 means that the ice is clean. A low albedo (ranging down to less than 0.1) implies contamination of the ice by carbonaceous or silicate dust particles and/or rock fragments; this could be because the outer part of the body accreted homogeneously and did not subsequently melt or differentiate in any way, or because the surface has picked up a coating of meteoritic dust over the ages. If the ice contains methane, it could also darken with age due to exposure to cosmic radiation. The grain size of the ice also has an effect, smaller grain sizes leading to a higher albedo.

asthenosphere A weak layer below the **lithosphere** of a terrestrial planet or icy moon, which behaves as a fluid on a geological time-scale, and which is capable of solid-state **convection** in response to the internal generation of heat at a rate sufficient to transport heat outwards as efficiently, or more efficiently, than conduction.

AU (astronomical unit) A convenient unit to measure interplanetary distances, defined as the mean distance of the Earth from the Sun, i.e. 150 million kilometres (93 million miles).

basalt A type of volcanic rock, with comparatively little silica in its make-up, and of low viscosity. The Earth's oceanic crust is formed of rocks of basaltic composition.

billion A thousand million (10^9).

caldera A crater formed largely by collapse of its floor, when an underlying magma body is removed, usually to supply lava flows or an explosive eruption.

carbonaceous chondrites Meteorites formed of **chondrules** in a matrix of hydrated minerals that also contains organic (carbon-bearing) molecules. Their bulk composition is thought to represent that of the non-volatile component of the **solar nebula**.

chondritic meteorites Primitive meteorites consisting of agglomerations of **chondrules**.

chondrule A silicate globule, about 1 mm to 1 cm in diameter commonly found in primitive meteorites. Believed to represent a droplet formed by condensation from the **solar nebula**.

clathrate Ice of one compound holding molecules of another at otherwise void spaces in its crystalline lattice. Ice I is capable of holding one molecule of methane (CH_4), carbon monoxide (CO), or nitrogen (N_2) for every six molecules of water (H_2O).

convection The transfer of heat in a fluid by means of circulation, which is driven by a strong temperature gradient.

crust Any outer, chemically and mineralogically distinct, part of the lithosphere of a planetary body, which overlies the **mantle**.

differentiation A gravity-driven process whereby a planet or satellite becomes layered. Denser phases segregate inwards and lighter phases work their way outwards.

eccentricity A measure of the departure of an orbit from circularity; the greater the eccentricity the more strongly pronounced the elliptical nature of the orbit.

ecliptic The plane of the Earth's orbit about the Sun. The orbits of most of the other planets are close to the same plane, the exception being Pluto, whose orbit is inclined at over 17°.

fumarole A vent from which volcanic gases escape, usually indicated by condensation around the vent (on the Earth, usually in the form of sulphur) and/or chemical alteration and discolouration of the surface material.

graben A valley formed by the down-dropping of its floor along steep faults that define the valley walls.

heterogeneous accretion Formation of planetary bodies by sequential condensation and **accretion**, beginning with the most **refractory** elements and progressing through progressively more **volatile** elements. If this happened, then the planetary bodies actually formed with a layered structure, although this is likely to have been modified by subsequent events.

homogeneous accretion A model for the formation of planetary bodies by the **accretion** of **refractory** and **volatile** elements at the same time. If this happened, then any layering within a planetary body has to be due to subsequent processes of **differentiation**.

horst An uplifted strip of terrain, bounded by faults on either side, or a block left standing high as a result of the down-dropping of **grabens** on either side.

igneous An adjective describing a rock that has previously been molten. If it was extruded on to the surface, then it is a volcanic igneous rock, otherwise it is intrusive.

ignimbrite A **pyroclastic** rock consisting of silica-rich fragments, often welded together and believed to be formed by the flow of rock fragments away from the site of an explosive eruption. The entrainment and heating of the surrounding air makes the flow buoyant and able to travel vast distances at high speed.

infra-red Radiation the wavelength of which is greater than that of visible light. Near-infra-red extends from 700 nm to about 3 μm in wavelength. The thermal infra-red is from about 3 μm to about 30 μm, and is the region where matter at normal temperatures emits radiation most strongly as a result of its heat.

isostasy   The operation of buoyancy forces on the outer, rigid layer of a planet. If this layer becomes thickened at the surface, the base of the layer will tend to subside if there is an underlying weak layer to accommodate the movement. The process stops when the weights per unit area of all columns of rock extending to the interior of the planet have become equal. Isostasy can operate only if the strength and thickness of the outer rigid layer are low enough.

lag deposit   A residue of non-volatile material on a surface, left when a more volatile component (with which the non-volatile component was originally mixed) has been removed.

late heavy bombardment   The earliest age of cratering of which good evidence remains (ending about 3.9 billion years ago), older traces having been obliterated by subsequent impacts. The late heavy bombardment appears to represent the tailing-off of the bombardment of the planets by debris left over from the formation of the solar system, as this debris became virtually exhausted. This event is well documented in the inner solar system by the radiometric dating of lunar samples, but it cannot be proven that the latest phase of heavy bombardment preserved within the satellite systems of each outer planet occurred at the same time, nor that it is the result of the same population of impactors.

limb   The sunlit edge of the visible disc of a **planetary body**.

liquidus   On a phase diagram of pressure against temperature, a line showing the conditions under which melting occurs.

lithosphere   The rigid outer layer of a terrestrial planet or icy moon, which overlies the **asthenosphere** (if present).

magma   Molten rock, especially when occurring in a body rising towards the surface (when erupted as a liquid it is known as lava). The term can be generalized to include melts within an icy moon.

magnetosphere   The region of space surrounding a planet in which the planet's magnetic field dominates over that of the solar wind.

mantle   The chemically distinct layer lying between the core and the **crust** of a planetary body. The mantle may be said to extend to the surface if there is no chemically distinct crust. Note that mantle and crust are chemical terms, whereas **lithosphere** and **asthenosphere** are defined by differences in physical properties.

moon   Used here to mean any natural satellite of a planet. The name of the Earth's satellite, the Moon, begins with a capital letter.

newtonian   Of a fluid, meaning that the rate at which it deforms is proportional to the applied shear stress divided by its viscosity. Non-newtonian fluids have a **yield strength** that must be overcome before they will deform.

nucleosynthesis   The creation of heavier elements by the fusion of lighter elements during nuclear fusion reactions within a star.

occultation When one astronomical object hides another by coming directly between it and the line of sight of the observer. See also **radio occultation**.

orbital resonance The situation when two bodies orbiting the same larger body have orbital periods that are a simple ratio of each other, for example Io, Europa, and Ganymede's $1:2:4$ orbital resonance about Jupiter, and Neptune and Pluto's $2:3$ orbital resonance about the Sun. Such situations are brought about by mutual gravitational and tidal interactions, and, once brought about, tend to be stable over long time periods.

palimpsest Usually a bright circular feature with little or no detectable relief, most common on Ganymede but also known on Callisto. Typically a hundred or more kilometres in diameter. Thought to represent the trace of a large impact that occurred when the **lithosphere** was still thin, the crater form having been destroyed either by viscous **relaxation** or the extrusion of warm ice or slush.

partial melting The phenomenon whereby a mixture of silicate minerals or a mixture of ices melts at a lower temperature than would any of its components were they in isolation. The initial melt has a different chemical composition to that of the solid, and (especially if it is less dense than the solid) may rise towards the surface to produce a volcanic eruption.

pixel A picture element in a digital image, the building block out of which an image is composed.

planetary body Used here to denote a planet or one of its satellites.

planetesimal An object from millimetres up to maybe a hundred kilometres in diameter formed during an intermediate stage of the formation of the solar system, by cohesion between the tiny grains that were formed by condensation from the **solar nebula**. Subsequently, the planets and satellites formed by gravitational **accretion** of planetesimals.

plate tectonics The generally accepted explanation of global **tectonics** on the Earth, which says that the **lithosphere** is divided into several rigid plates. These are spreading apart at rates of up to about 10 cm per year at oceanic spreading axes (mid-ocean ridges), and converging at continental collision zones and subduction zones beneath island arcs and Andes-type mountain ranges where lithosphere is destroyed at a rate sufficient to balance its creation at spreading axes.

population I Originally defined at Saturn, and applied also to Uranus; referring to the oldest suite of craters preserved in the satellite system (and the impactors that produced these). Generally thought to represent the sweeping up of post-accretional debris, although not necessarily the same as the **late heavy bombardment**.

population II As for **population I**, but a later phase of cratering, producing a smaller proportion of large craters. Possibly the result of collisions with debris in orbit around the same planet. Such debris could have resulted from the destruction of a satellite because of a major impact. Satellites that preserve reasonable evidence of geological activity generally have terrains, or even their entire surface, that were **resurfaced** after the end of the population I bombardment, and are dominated by population II craters alone. The population II flurry of impacts is unlikely to date from the same time at each planet.

population III The gradually declining background flux of impacts (by cometary material and other debris) continuing to the present day, which is sometimes distinguished from **population** II (which is thought to represent one or more shorter flurries).

prograde Referring to the orbital motion of a planet when it is anticlockwise as seen from above the plane of the solar system (above the Sun's north pole) and, similarly, of a satellite when its orbit is anticlockwise as seen from above its planet's north pole. Also used for rotation of a planet or satellite on its axis that is anticlockwise as seen from above the body's north pole. Most bodies in the solar system have both prograde orbits and prograde rotation. The opposite is **retrograde**.

protoplanet In models for solar system formation and development, a large **planetesimal** that, because of its larger gravity field, grows faster than the others in its vicinity. Eventually it reaches planetary size, **differentiates** and becomes a planet.

protosatellite disc A disc of gas and dust around a planet, or **protoplanet**, from which its satellites formed.

pyroclastic A term describing a volcanic eruption and its resulting deposits when it is caused by explosive fragmentation of the molten rock (or ice) as a result of the pressure built up by gases escaping from or through the melt.

radio occultation A technique used to determine a profile of the density of the atmosphere with respect to altitude for a planetary body, by measuring the way in which a radio signal from a space probe is weakened as the signal passes through the atmosphere while the space probe passes behind the planetary body as seen from Earth. These density data give information on the ratio between temperature and the mean molecular weight of the atmosphere, so if temperature can be determined independently, say by **infra-red** techniques, radio occultation data can be used to determine the mean molecular weight of the atmosphere.

radiogenic heating Heating of a planetary body from within by heat generated in the decay of radioactive elements. The short-lived isotope ^{26}Al was probably very important during the first few million years of the solar system's life, but subsequently most radiogenic heat has originated in much longer-lived isotopes of uranium, thorium, and potassium. All these elements reside within silicates, and there is no significant source of radiogenic heat within ices.

radiometric dating A method of working out the time since a rock or mineral grain formed by measuring the isotopes of elements involved in radioactive decay.

refractory A relative term, denoting the earliest substances to condense from a cooling vapour (as in the **solar nebula**), while the temperature was still high.

regolith A layer of fragmentary debris on the surface of a solid body, produced by meteoritic impacts.

relaxation The process whereby a **lithosphere** deforms under its own weight, in response to major topographic features such as large craters or **grabens**. The result is not a flat surface, but one on which the relief is subdued. Occurs in lithospheres that are thin and/or warm.

reseau marks An array of calibration dots on the faceplate of a vidicon camera, which are put there to enable electronic geometric distortions to be corrected. If reseau marks are not synthetically removed from an image they appear as black points.

resolution A measure of the size of the smallest detail visible on an image. In this book the values of resolution quoted represent the size of the pixels; elsewhere resolution is sometimes quoted in terms of km per line pair.

resurfacing Creation of a new surface on a planetary body by **volcanic** or **tectonic** processes. These can include burial of older features under the products of eruptions and the break-up of old **terrains** by faulting.

retrograde Referring to the orbital motion of a satellite if it is clockwise as seen from above its planet's north pole; the orbits of most bodies in the solar system have the opposite, or prograde, sense. Also used for rotation of a planet on its axis that is clockwise when seen from north of the plane of the **ecliptic**.

rheological Pertaining to the way a material responds to stress applied at different rates, in particular how it can flow in the solid state under high pressure.

shear strain The amount by which a fluid (or any other substance) is deformed in response to a **shear stress**.

shear stress The force tending to deform something by shearing one layer over another.

silicate A silicate mineral is any of the common rock-forming minerals on the Earth and the other terrestrial planets, consisting of silicon and oxygen, usually with a mixture of any of iron, magnesium, sodium, potassium, calcium, or aluminium. 'Silicates' is often used as a term to mean minerals or rocks in general.

solar nebula The cloud of gas and dust around the proto-Sun from which the solar system formed.

sputtering The vaporization of ice as a result of impact by micrometeorites or energetic cosmic ray particles. This proceeds at extremely slow rates on the icy satellites, but it could help to increase the concentration of rock fragments on the surface. See also **sublimation**.

strain rate The rate at which a material is made to deform.

strike-slip A geological term referring to faults in which the dominant movement is lateral, so as to displace features sideways from their original positions.

sublimation The process whereby ice, in a vacuum, evaporates directly from the solid to the vapour state. This occurs extremely slowly at the low temperatures prevalent in the outer solar system, but over the age of the solar system it could provide, by removal of the ice, a possible mechanism for concentrating a residual **lag deposit** of rocky material on the surface of an icy satellite. See also **sputtering**.

synchronous rotation Rotation of a planetary satellite in the same period as its orbit, so it always keeps the same face towards the planet. This is brought about through tidal interactions and is also known as captured rotation.

tectonics A geological term referring to processes of faulting or other distortion or disruption of a body's **lithosphere**, often on a global scale, and almost always as a result of large-scale internal movements below the lithosphere.

terminator The boundary between the sunlit and dark parts of the surface of a **planetary body**.

terrain A tract of surface that has a set of recognizable characteristics, indicating that it probably has a different history to that of other terrains on the same body.

terrestrial Referring to something that occurs on the Earth. Also, in the term terrestrial planet, meaning any of the planets whose outer layers are dominated by silicates, i.e. Mercury, Venus, Earth, the Moon, and Mars (and, sometimes, Io).

tidal heating (tidal dissipation) Heating of a satellite as a result of varying deformation due to the action of its planet's gravity on a tidal bulge.

transform fault In the Earth's oceans a sideways offset between two segments of a spreading axis. As a result of the axial spreading, a transform fault across which the spreading axis is offset to the right is actually a site of relative motion to the left.

transcurrent fault A simple **strike-slip** fault, in which the relative motion between units on either side is the same as the visible sense of displacement between features offset by the fault (compare **transform fault**).

viscosity A measure of the ease with which a fluid flows; it can be defined as **shear stress** divided by rate of **shear strain**. Conventionally measured in poise (1 poise $= 0.1$ kg m^{-1} s^{-1}). Typical values of viscosity are: water, 10^{-2} poise; honey and $NH_3 \cdot 2H_2O$ (ammonia hydrate melt), 100 poise; sulphur, about 0.1 to nearly 10^3 poise according to temperature; terrestrial basalt lava, 10^3 poise; terrestrial rhyolite lava, 10^6 poise; and pure ice at 240 K, 10^{16} poise.

viscous relaxation see **relaxation**.

volatile In cosmochemistry a relative term, denoting substances that do not condense from a cooling vapour (as in the **solar nebula**) until the temperature has dropped very low. Used in volcanology to denote an abundant chemical species in the vapour phase that escapes from **magmas**, and may cause an explosive (**pyroclastic**) eruption.

volcanism The eruption of molten material or gas-driven solid fragments at the surface of a **planetary body**. On icy moons, the phenomenon is sometimes called cryovolcanism to distinguish it from silicate or sulphur volcanism.

yield strength The property of a non-newtonian fluid, such that a minimum **shear stress** must be applied to it before it will begin to flow. Terrestrial lavas and icy **magmas** all have yield strengths.

Bibliography

I wrote this book because I could not find one like it at a similar level. The ones to consult if you want a broad overview of planetary satellites as a whole, with considerable technical detail, are:

Burns, J. A. and Matthews, M. S. (ed.) (1986). *Satellites*, Space Science Series, 1021pp. University of Arizona Press, Tucson.

and the still-useful:

Morrison, D. (ed.) (1982). *Satellites of Jupiter*, Space Science Series, 972pp. University of Arizona Press, Tucson.

Two books at a generally less detailed level than this one, that cover geology on all the terrestrial planets and, briefly, their moons are:

Greeley, R. (1994). *Planetary landscapes*, 2nd edn. 275pp. Chapman and Hall, London.
Morrison, D. and Owen, T. (1988). *The planetary system*, 519pp. Addison-Wesley, Reading, Massachusetts.

An up-to-date and easily readable review of the whole solar system is:

Beatty, J. K. and Chaikin, A. (1990). *The new solar system*, 3rd edn., 326pp. Sky, Cambridge, Massachusetts, and Cambridge University Press, Cambridge.

The techniques used in mapping planetary landscapes and the conventions used for naming features are discussed at a simple level in:

Greeley, R. and Batson, R. M. (ed.) (1990). *Planetary mapping*, Cambridge Planetary Science Series 6, 296pp. Cambridge University Press, Cambridge.

Accretion models and ideas on the chemical development of solar system material are fast evolving. A recent, rather technical, compilation is:

Weaver, H. A. and Danly, L. (ed.) (1989). *The formation and evolution of planetary systems*, Space Telescope Science Institute Symposium Series 3, 344pp. Cambridge University Press, Cambridge.

In writing this book, I have made use of material from several of the above, most especially of several of the chapters in the volume edited by Burns and Matthews (1986). For those who want to find their way into the rest of the literature on the formation and geological evolution of the planetary satellites covered in this book, I give below a list of the other principal references, including the initial reports of the *Voyager* encounters with each planetary satellite system (there are usually other papers in the same journal issue that I have not listed), a few seminal papers, and the more useful papers of other sorts. It is not an exhaustive bibliography, and it concentrates on works that post-date Burns and Matthews (1986). I have not included any conference proceedings and the like, although I found the abstract volumes for the Lunar and Planetary Science Conference (LPSC) held annually at the Lunar and Planetary Institute, Houston, very useful.

Allison, M. L. and Clifford, S. M. (1987). Ice-covered water volcanism on Ganymede. *Journal of Geophysical Research*, **92**, 7865–76.

Binzel, R. P. (1990). Pluto. *Scientific American*, June 1990, 26–33.

Cameron, A. G. W. (1988). Origin of the solar system. *Annual Review of Astronomy and Astrophysics*, **26**, 441–72.

Consolmagno, G. J. (1985). Resurfacing Saturn's satellites: models of partial differentiation and expansion. *Icarus*, **64**, 401–13.

Consolmagno, G. J. and Lewis, J. S. (1978). The evolution of icy satellite interiors and surfaces. *Icarus*, **34**, 280–93.

Crater Analysis Techniques Working Group (1979). Standard techniques for presentation and analysis of crater size–frequency data. *Icarus*, **37**, 467–74.

Croft, S. K. and Soderblom, L. A. (1991). Geology of the Uranian satellites. In *Uranus* (ed. J. T. Bergstralh, E. D. Miner, and M. S. Matthews), pp. 561–628. University of Arizona Press, Tucson.

Croft, S. K., Lunine, J. I., and Kargel, J. (1988). Equation of state of ammonia–water liquid: derivation and planetological applications. *Icarus*, **73**, 279–93.

Ellsworth, K. and Schubert, G. (1983). Saturn's icy satellites: thermal and structural models. *Icarus*, **54**, 490–510.

Golombek, M. P. and Banerdt, W. B. (1990). Constraints on the subsurface structure of Europa. *Icarus*, **83**, 441–52.

Hapke, B. (1989). The surface of Io: a new model. *Icarus*, **79**, 56–74.

Herrick, D. L. and Stevenson, D. J. (1990). Extensional and compressional instabilities in icy satellite lithospheres. *Icarus*, **85**, 191–204.

Ingersoll, A. P. (1990). Dynamics of Triton's atmosphere. *Nature (London)*, **344**, 315–17.

Janes, D. M. and Melosh, H. J. (1988). Sinker tectonics: an approach to the surface of Miranda. *Journal of Geophysical Research*, **93**, 3127–43.

Jankowski, D. G. and Squyres, S. W. (1988). Solid-state ice volcanism on the satellites of Uranus. *Science*, **241**, 1322–5.

Johnson, M. W. and Nicol, M. (1987). The ammonia–water phase diagram and its implications for icy satellites. *Journal of Geophysical Research*, **92**, 6339–49.

Johnson, T. V. and Soderblom, L. A. (1983). Io. *Scientific American*, December 1983, 60–71.

Johnson, T. V., Brown, R. H., and Pollack, J. B. (1987). Uranus satellites: densities and composition. *Journal of Geophysical Research*, **92**, 14884–94.

Johnson, T. V., Veeder, G. J., Matson, D. L., Brown, R. H., Nelson, R. M., and Morrison, D. (1988). Io: evidence for silicate volcanism in 1986. *Science*, **242**, 1280–3.

Kargel, J. S., Croft, S. K., Lunine, J. I., and Lewis, J. S. (1991). Rheological properties of ammonia–water liquids and crystal–liquid slurries: planetological applications. *Icarus*, **89**, 93–112.

Lewis, J. S. (1971). Satellites of the outer planets: their physical and chemical nature. *Icarus*, **15**, 174–85.

Lindal, G. F., Wood, G. E., Hotz, H. B., Sweetman, D. N., Eshleman, V. R., and Tyler, G. L. (1983). The atmosphere of Titan: an analysis of Voyager-1 radio occultation measurements. *Icarus*, **53**, 348–63.

Lucchitta, B. K. (1980). Grooved terrain on Ganymede. *Icarus*, **44**, 481–501.

Lunine, J. I. (1989). Sulfur lakes and silicate flows: thermal emissions from Io's hot spots. *Time-variable phenomena in the Jovian system*, NASA SP–494 (ed. J. S. Belton, R. A. West, and J. Rahe), pp. 63–70. NASA, Washington DC.

Lunine, J. I. (1989). The Urey Prize Lecture: Volatile processes in the outer solar system. *Icarus*, **81**, 1–13.

Lunine, J. I. and Stevenson, D. J. (1982). Formation of the galilean satellites in a gaseous nebula. *Icarus*, **52**, 14–39.

Lunine, J. I. and Stevenson, D. J. (1985). Physics and chemistry of sulphur lakes on Io. *Icarus*, **64**, 345–67.

Masursky, H., Schaber, G. G., Soderblom, L. A., and Strom, R. G. (1979). Preliminary geological mapping of Io. *Nature* (*London*), **280**, 725–33.

McEwen, A. S. and Soderblom, L. A. (1983). Two classes of volcanic plumes on Io. *Icarus*, **55**, 191–217.

McEwen, A. S., Lunine, J. I., and Carr, M. H. (1989). Dynamic geophysics of Io. *Time-variable phenomena in the Jovian system*, NASA SP–494 (ed. J. S. Belton, R. A. West, and J. Rahe), pp. 11–46. NASA, Washington DC.

McKinnon, W. B. and Melosh, H. J. (1980). Evolution of planetary lithospheres: evidence from multiringed structures on Ganymede and Callisto. *Icarus*, **44**, 454–71.

Moore, J. M. (1984). The tectonic and volcanic history of Dione. *Icarus*, **59**, 205–20.

Moore, J. M. and Spencer, J. R. (1990). Koyaanismuuyaw: the hypothesis of a perennially dichotomous Triton. *Geophysical Research Letters*, **17**, 1757–60.

Muhleman, D. O., Grossman, A. W., Bulter, B. J., and Slade, M. A. (1990). Radar reflectivity of Titan. *Science*, **248**, 975–80.

Owen, T. (1982). Titan. *Scientific American*, February 1982, 76–85.

Parmentier, E. M., Squyres, S. W., Head, J. W., and Allison, M. L. (1982). The tectonics of Ganymede. *Nature* (*London*), **295**, 290–3.

Passey, Q. R. (1983). Viscosity of the lithosphere of Enceladus. *Icarus*, **53**, 105–20.

Peale, S. J., Cassen, P., and Reynolds, R. T. (1979). Melting of Io by tidal dissipation. *Science*, **203**, 892–4.

Pieri, D. C., Baloga, S. M., Nelson, R. M. and Sagan, C. (1984). Sulphur flows of Ra Patera, Io. *Icarus*, **60**, 685–700.

Plescia, J. B. (1983). The geology of Dione. *Icarns*, **56**, 255–77.

Plescia, J. B. (1987). Geological terrains and crater frequencies on Ariel. *Nature* (*London*), **327**, 201–4.

Plescia, J. B. (1988). Cratering history of Miranda: implications for geologic processes. *Icarus*, **73**, 442–61.

Plescia, J. B. and Boyce, J. M. (1982). Crater densities and geological histories of Rhea, Dione, Mimas and Tethys. *Nature* (*London*), **295**, 285–90.

Plescia, B. J. and Boyce, J. M. (1983). Crater numbers and geological histories of Iapetus, Enceladus, Tethys and Hyperion. *Nature* (*London*), **301**, 666–70.

Poirier, J. P. (1982). Rheology of ices: a key to the tectonics of the ice moons of Jupiter and Saturn. *Nature* (*London*), **299**, 683–8.

Ross, M. N. and Schubert, G. (1985). Tidally forced viscous heating in a partially molten Io. *Icarus*, **64**, 391–400.

Ross, M. N. and Schubert, G. (1987). Tidal heating and an internal ocean model of Europa. *Nature* (*London*), **325**, 133–4.

Ross, M. N. and Schubert, G. (1990). The coupled orbital and thermal evolution of Triton. *Geophysical Research Letters*, **17**, 1749–52.

Ross, M. N., Schubert, G., Spohn, T., and Gaskell, R. W. (1990). Internal structure of Io and the global distribution of its topography. *Icarus*, **85**, 309–25.

Sagan, C. and Chyba, C. (1990). Triton's streaks as windblown dust. *Nature* (*London),* **346**, 546–8.

Schaber, G. G. (1980). The surface of Io: geologic units, morphology and tectonics. *Icarus*, **43**, 302–33.

Schenk, P. M. (1989). Crater formation and modification on the icy satellites of Uranus and Saturn: depth/diameter and central peak occurrence. *Journal of Geophysical Research*, **94**, 3813–32.

Schenk, P. M. (1991). Fluid volcanism on Miranda and Ariel: flow morphology and composition. *Journal of Geophysical Research*, **96**, 1887–906.

Schenk, P. M. (1991). Ganymede and Callisto: complex crater formation and planetary crusts. *Journal of Geophysical Research*, **96**, 15635–64.

Schenk, P. M. and McKinnon, W. B. (1987). Ring geometry on Ganymede and Callisto. *Icarus*, **72**, 209–34.

Schenk, P. M. and McKinnon, W. B. (1989). Fault offsets and lateral crustal movement on Europa: Evidence for a mobile ice shell. *Icarus*, **79**, 75–100.

Simonelli, D. P., Pollack, J. B., McKay, C. P., Reynolds, P. T., and Summers, A. (1989). The carbon budget in the outer solar nebula. *Icarus*, **82**, 1–35.

Sinton, W. M. and Kaminski, C. (1988). Infrared observations of eclipses of Io, its thermophysical parameters, and the thermal radiation of the Loki volcano and environs. *Icarus*, **75**, 207–32.

Smith, B. A., Soderblom, L. A., Beebe, R., Boyce, J., Briggs, G., Carr, M. *et al.* (1979). The galilean satellites and Jupiter: Voyager 2 imaging science results. *Science*, **206**, 927–50.

Smith, B. A., Soderblom, L. A., Johnson, T. V., Ingersoll, A. P., Collins, S. A., Hunt, G. E. *et al.* (1979). The Jupiter system through the eyes of Voyager 1. *Science*, **204**, 951–71.

Smith, B. A., Soderblom, L., Beebe, R., Boyce, J., Briggs, G., Bunker, A. *et al.* (1981). Encounter with Saturn: Voyager 1 imaging science results. *Science*, **212**, 163–91.

Smith, B. A., Soderblom, L. A., Beebe, R., Bliss, D., Boyce, J., Brahic, A. *et al.* (1986). Voyager 2 in the Uranian system: Imaging science results. *Science*, **233**, 43–64.

Smith, B. A., Soderblom, L., Banfield, D., Barnet, C., Basilevsky, A. T., Beebe, R. F. *et al.* (1989). Voyager 2 at Neptune: imaging science results. *Science*, **246**, 1422–49.

Soderblom, L. A. and Johnson, T. V. (1982). The moons of Saturn. *Scientific American*, January 1982, 73–86.

Soderblom, L. A., Kieffer, S. W., Becker, T. L., Brown, R. H., Cook, A. F. II, Hansen C. J., Johnson, T. V., Kirk, R. L., and Shoemaker, E. M. (1990). Triton's geyser-like plumes: discovery and basic characterization. *Science*, **250**, 410–15.

Spencer, J. R. (1990). Nitrogen frost migration on Triton: a historical model. *Geophysical Research Letters*, **17**, 1769–72.

Spencer, J. R. and Maloney, P. R. (1984). Mobility of water ice on Callisto: evidence and implications. *Geophysical Research Letters*, **11**, 1223–6.

Spencer, J. R., Shure, M. A., Ressler, M. E., Goguen, J. D., Sinton, W. M., Toomey, D. W., Denault, A., and Westfall, J. (1990). Discovery of hotspots on Io using disk-resolved infrared imaging. *Nature* (*London*), **348**, 618–21.

Squyres, S. W., Reynolds, R. T., Cassen, P. M., and Peale, S. J. (1983). The evolution of Enceladus. *Icarus*, **53**, 319–31.

Stansbury, J. A., Lunine, J. I., Porco, C. C., and McEwen, A. S. (1990). Zonally averaged thermal balance and stability models for nitrogen polar caps on Triton. *Geophysical Research Letters*, **17**, 1773–6.

Stern, S. A. (1989). Pluto: comments on crustal composition, evidence for global differentiation. *Icarus*, **81**, 14–23.

Stevenson, D. J. (1982). Volcanism and igneous processes in small icy satellites. *Nature* (*London*), **298**, 142–4.

Stevenson, D. J. and Lunine, J. I. (1986). Mobilization of cryogenic ice in outer Solar System satellites. *Nature* (*London*), **323**, 142–4.

Strom, R. G., Terrile, R. J., Masursky, H., and Hansen, C. (1979). Volcanic eruption plumes on Io. *Nature (London)*, **280**, 46–8.

Thomas, P. J. and Squyres, S. W. (1988). Relaxation of impact basins on icy satellites. *Journal of Geophysical Research*, **93**, 14919–32.

Thomas, P. J. and Squyres, S. W. (1990). Formation of crater palimpsests on Ganymede. *Journal of Geophysical Research*, **95**, 19161–74.

Trafton, L. M. (1990). Pluto's atmosphere at perihelion. *Geophysical Research Letters*, **16**, 1209–13.

Wood, J. A. (1988). Chondritic meteorites and the solar nebula. *Annual Review of Earth and Planetary Science*, **16**, 53–72.

Young, A. T. (1984). No sulphur flows on Io. *Icarus*, **58**, 197–226.

Index

Plain numbers refer to a concept in the text or (rarely) in a table on the referenced page. Numbers in italics refer to figures or their captions, or (where noted) to colour plates. The names of specific surface features are indexed alphabetically under the name of the satellite or other body on which they occur. Note that: many terms are defined in the Glossary on pages 189–95; the important physical parameters for the terrestrial planets and all satellites greater than about 200 km in radius are in given Table 1.1 (page 4); and the sizes of all known satellites of the outer planets are compared in Figure 1.2 (page 5).